Lionel Boudouin Tamtchoun
Abdoul Wahabou
Prosper Samba Koukouare

DRILLING ENGINEERING

Lionel Boudouin Tamtchoun
Abdoul Wahabou
Prosper Samba Koukouare

DRILLING ENGINEERING

Oil drilling design and planning

ScienciaScripts

Imprint

Cover image: www.ingimage.com

This book is a translation from the original published under ISBN 978-3-8416-1432-2.

Publisher:
Sciencia Scripts
is a trademark of
Dodo Books Indian Ocean Ltd. and OmniScriptum S.R.L publishing group

120 High Road, East Finchley, London, N2 9ED, United Kingdom
Str. Armeneasca 28/1, office 1, Chisinau MD-2012, Republic of Moldova, Europe
Managing Directors: Ieva Konstantinova, Victoria Ursu
info@omniscriptum.com

Printed at: see last page
ISBN: 978-620-8-39694-7

TABLE OF CONTENTS

SUMMARY

This document describes the drilling programme for the Dokety-1 (A4-1) well. This is a deviated exploration well in the Doba Basin in Chad whose objective is to intersect 2 targets: Kedeni and Mangara located at depths of 9576.44 ft and 9037.4 ft respectively. In order to carry out this study efficiently, Halliburton Landmark and E-RedBook drilling engineering software were used to design the drilling programme. The work was carried out following a precise design process: Firstly, we designed the drilling trajectory using COMPASS™ software, then the casing programme was developed using StressCheck™ and CasingSeat™ software, taking into account all possible worst-case scenarios that could destabilise the casing, including water hammer for burst strength and loss of drilling fluid for crush strength. This was followed by the design of the cementing programme, drilling mud and drilling tool. Finally, a detailed estimate of the drilling time and cost was made using WellCost™ software. A "Build and Hold" type of deviated drilling was obtained. It will be carried out in 4 phases: Conductor beating (diameter 20", grade H-40, shoe depth 213.36 ft), surface section (diameter 13 3/8", grade K-55, shoe depth 2,296.6 ft, TMT tool 17 1/2"), intermediate section (diameter 9 5/8", grade N-80, shoe depth 8,208.7 ft, PDC tool 12 1/4") and production section (diameter 7", grade N-80, shoe depth 11,056.4 ft, PDC tool 8 1/2"). In addition, other hardware components of the drilling system have been sized to ensure good rig performance: BOP operating pressure (5,200 psi), a 1,290.6 HP winch, 3 1,102.7 HP mud pumps, 1,456.89 HP rotation power and a 45.5 m high derrick. Finally, an estimate of the duration and cost of drilling was carried out. This showed that drilling the Dokety-1 well would take 33.39 days at a total cost of **$6**,150,216.

Key words: Tubing, shoe, grade, tool, deviated drilling, mud.

GENERAL INTRODUCTION

From the start of the black gold rush in 1859 to the present day, demand for fossil fuels has continued to rise (Lavoisy, 2022). With reserves being depleted at a rapid rate day by day, oil companies and governments routinely set up exploration programmes with the aim of finding new reserves to meet the ever-increasing demand for fossil fuels. With this in mind, the Chinese National Petroleum Company (CNCP), in partnership with the State of Chad, has set up an onshore drilling project in the Logone Oriental region of Chad. The project involves the construction of a deviated exploration well in the Doba basin. This operation requires a major investment and presents risks for the personnel present on site, which is why an in-depth study and meticulous planning are of major importance in order to ensure the safety of the personnel and to complete the work at minimum cost.

The environment in which an oil well is to be constructed is a real challenge, given the many constraints to be faced and the multitude of variables to be taken into account in order to achieve the targeted objectives safely and at the lowest possible cost. A lack of planning, or a failure to plan properly, would therefore be a handicap to operations, and could lead to a lack of safety for the personnel present on site, a delay in completion and abnormally high costs.

The main objective of this investigation is to design a detailed drilling programme for a deviated well capable of reaching the target safely, and to establish the project budget.

In order to achieve this objective, specific targets have been set:

- Optimal selection of rig and main drilling equipment;
- Designing a suitable mud programme ;
- Establish a casing and cementing programme in line with the project objectives;
- Planning the drilling trajectory ;
- Evaluate the duration and overall cost of the project and draw up a budget.

This document is structured in three chapters with a general introduction and a general conclusion. The first chapter deals with the general aspects of designing a drilling programme. The second chapter deals with the equipment and methods used to carry out this work, and finally the third chapter deals with the results obtained from this work and their interpretation.

Chapter I: GENERAL

Introduction

Well planning is an area of drilling engineering which, based on numerous variables, formulates a drilling programme with a certain number of characteristics (safety, minimum cost, long life of the structure). This operation is one of the most demanding in drilling engineering. It requires the integration of engineering principles, business philosophy and experience factors (Neal, 2006). Although planning methods and practices may vary within the drilling industry, the end result must be a well drilled safely and at minimum cost that meets the requirements of reservoir and production engineers. This chapter will therefore present in detail each of the mechanisms that make up a modern drilling system, as well as the general method of designing a drilling programme for a deviated well.

I.1 Definition and principle of drilling

Drilling is a technique used in various fields of engineering (oil and gas engineering, civil engineering, mining engineering, etc.). In the oil and gas sector, this involves digging a well into the ground using a drill string, with the aim of reaching porous and permeable rocks in the subsoil that are likely to contain liquid or gaseous hydrocarbons. In the final stages before production, this technique removes or reduces many uncertainties about the prospect, whether hydrocarbons are present, their nature and the volume of reserves. However, questions may remain about profitability, the shape of the deposit and the homogeneity of its characteristics. It is therefore necessary to drill several wells at different locations in the reservoir, to better delineate the deposit and choose the best locations for future production wells. These additional wells make up the appraisal programme, at the end of which a decision is made whether to exploit the deposit or abandon it. (Azar, 2020).

I.2 Drilling platform systems

Although there are different types of drilling rig, they share many similarities in that the tasks they perform are identical. Five systems are typically found on modern drilling rigs: the power system, the lift system, the rotary system, the circulation system and the well control system. Figure 1 shows the main components of a drilling rig (King, 2020).

1	The crown block
2	The mast
3	The monkey board
4	The mobile block
5	The hook
6	The pivot
7	The rotating hose
8	La kelly
9	Kelly bushing
10	Master bushing
11	The mouse hole
12	The rat hole
13	The lifting winch
14	The weight indicator
15	The drilling console
16	The control station
17	The rotating hose
18	The battery
19	The footbridge
20	The stem ramp
21	The rod rack
22	The substructure
23	The mud return line
24	The vibrating sieve
25	The choke manifold
26	The sludge gas separator
27	The deaerator
28	The reserve pit
29	Sludge pits
30	The desander
31	The desiccant
32	Sludge pumps
33	The sludge discharge line
34	Storage of waste sludge
35	The mud house
36	The water tank
37	Fuel storage
38	Motors and generators
39	The drilling line

Figure 1The components of a drilling platform (King, 2020)

- **The feed system**

It supplies power to the other main systems on the rig and to other auxiliary systems such as pumps, motors, etc. The mode of power transmission on the rig can be mechanical, direct current (DC) electrical or alternating current (AC) electrical, (King, 2020).

- **The circulation system**

This is the system that allows the drilling fluid or mud to circulate down through the hollow drill string and up through the annulus between the drill string and the wellbore. It is a continuous system of pumps, distribution lines, storage tanks, storage pits and cleaning units that allows the drilling fluid to achieve its main objectives (King, 2020).

- **The rotary system**

This is the system that rotates the drill string and therefore the drill bit at the bottom of the borehole. This function is provided by a rotation table (in a Kelly system) or a top-drive. We note that the functions of the kelly, kelly bushing, master bushing, and rotation table on a rig with a top drive are provided by the top drive (King, 2020).

- **The lifting system**

This system does most of the work on the platform. It is used to raise, lower and suspend the drill string, and to suspend the casing column using a winch (Drawworks).

- **The well control system or blowout prevention system**

This is the system that prevents the uncontrolled and catastrophic release of high-pressure fluids (oil, gas or water) from underground formations. These uncontrolled releases of formation fluids are called blowouts. The system essentially consists of a BOP and pressurised fluid accumulators. The BOP is the main component of the well control system. It is hydraulically operated. Pressurised fluids are used to operate jacks and cylinders to open or close the shut-off valves. Accumulators are used to store a non-explosive gas under pressure and hydraulic fluids to operate the hydraulic systems on the platform. (King, 2020).

I.3 The components of a drilling programme

A drilling programme is a structured set of information describing the procedure to be followed and the values of the various variables to be taken into account when drilling. It can be subdivided into 5 main components: The mud programme, the casing programme, the cementing programme, the drilling tool programme and the well control programme.

I.3.1 Drilling mud

Drilling fluid, also known as drilling mud, is a mixture of water, oil, clay and various chemicals. This fluid fulfils a number of functions, including transporting cuttings, cooling the drill bit and drill string, controlling formation pressure, etc. Because of its density, drilling mud is able to exert hydrostatic pressure on the well walls to prevent fluid from coming in. (Faridah, 2021). This pressure can be higher than that of the formation fluids (overbalanced drilling) or lower (underbalanced drilling). There is a range of mud pressure for which the well walls will remain stable during drilling: this is the "mud window". It is characterised by a lower and an upper limit that depend on the mechanical stability (mechanical stability of the well, in-situ stresses and the pore pressure) of the well as well as various other technical requirements (Wolfgang, 2016).

I.3.2 Tubing

The cost of casing represents a significant proportion of the drilling budget. Therefore, proper planning of casing installation depths and casing selection is essential to achieve a safe and cost effective well. Casing design must take into account the effects of pressure and temperature changes that can occur at any time or depth during drilling or well operation. For each of the stress regimes, calculations must be made to establish that there is an adequate margin of resistance in the casing string at all depths. Information required for casing design includes: mud weights, formation pressures, fracturing gradients, casing shoe depths, drilling direction, cementing programme, and temperature profile. A casing string is generally made up of (Kilian, 2007) :

- **A conductive tube:** 20 to 36" in diameter, its role is to protect unconsolidated surface formations, seal off areas of shallow water and protect the platform foundation.
- **Surface casing:** These vary in diameter from 24" to 17" and their function is to prevent the collapse of loose formations encountered at shallow depths and to isolate the water table. It should be noted that the depth at which surface casing is laid is often determined by government or company policy and is not always selected on technical grounds.
- **Intermediate casing:** This casing string is purely technical, as it must be able to control abnormal formation pressures, swelling clay formations, circulation losses or collapse zones. Intermediate casing diameters range from 9" to 17 1/2". This casing supports the BOP and later the final production wellhead.
- **Production casing:** This has a diameter of between 5" and 9 5/8" and is used to isolate the production zones and ensure control of the reservoir fluid.
- **The liner:** To reduce costs, the installed production casing sometimes does not reach the surface but ends in the previous section. This type of casing configuration is called a liner. This is mounted on a "liner hanger" installed on the previous casing column.

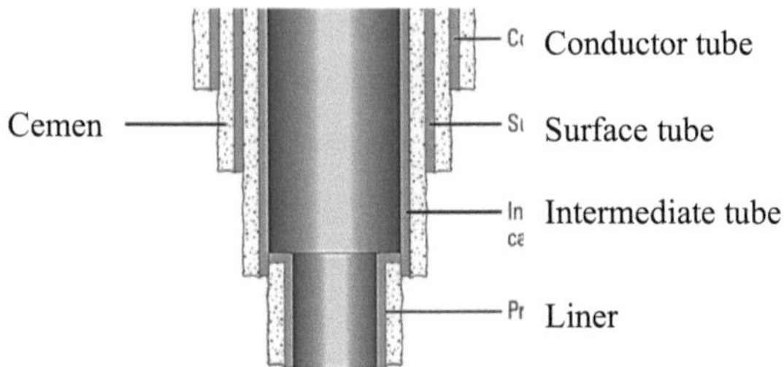

Figure 2 Structure of a cased and cemented well (Nelson, 2011)

I.3.3 Sizing the casing

Sizing the casing is a complex operation that involves selecting tubes with appropriate characteristics to ensure the long-term integrity of the well. This operation is carried out in 3 main stages: Determining the shoe depth, selecting the pipe diameter and selecting the grade of steel for each casing section.

I.3.3.1 Pipe-laying depths

In general, the calculation of casing installation depths begins at the bottom of the well. After determining the dimensions of the hole to be drilled and applying the corresponding mud weight, a kick is assumed and the casing installation depth is calculated for a kick pressure that can fracture the formation. As mentioned previously, the surface casing installation depth is normally determined by government or local regulations. Until now, the determination of casing installation depths has been based solely on the fracture gradients of the different formations and the mud weights of the different sections. The formations to be drilled also influence the determination of the casing installation depth. In practice, a casing is normally placed in a competent formation (Kilian, 2007).

I.3.3.2 Selection of casing diameter

The casing diameter depends on the properties of the fluids likely to be encountered in the reservoir and the flow rate. The choice of casing diameter is generally made starting from the narrowest section of the well towards the upper sections. The first step is to determine the diameter of the production casing and then deduce the diameter of the other sections based on the API standard standard (Wolfgang, 2016).

I.3.3.3 Selection of the casing grade

The casing grade defined by API corresponds to the quality of the steel and is a function of its mechanical strength. In order to eliminate any risk of casing failure, it is vital to assess the loads to which casings may be subjected (bursting, crushing, tension) in order to make an optimum choice of grade. To calculate the burst and collapse pressure for which the casing should be designed, the differential pressure (external pressure - internal pressure) is determined assuming worst-case scenarios (Wolfgang, 2016).

- **Determining burst pressure**

To determine the burst pressure, we assume a "kick" or water hammer. This means that the highest burst pressure should be at the top of the casing and the lowest at the shoes (because the hydrostatic pressure in the annulus counterbalances the burst pressure).

$$P_b = P_f . F \quad (1)$$

With :

P_b : burst pressure (psi) ;
P_f Maximum anticipated formation pressure for drilling the next section (psi) ;
F: Safety factor.

- **Determining crushing pressure**

To determine the crushing pressure, it is assumed that the mud inside the casing is lost in a formation below in a fractured formation. The crushing pressure is therefore due to the hydrostatic pressure of the fluid outside the casing and is therefore maximum at the casing shoe and zero at the casing head.

$$P_c = 0.052 . \rho_m . D . F \quad (2)$$

With :

P_c crushing pressure ;
D: Depth ;
ρ_m density of the drilling mud ;
F: Safety factor.

- **Determining the axial load**

The tensile forces acting on the casing are due to its weight, bending forces and impacts during landing. It should be noted that for highly deviated wells, landing the casing is only possible when it is partially or totally empty. Here, the casing is closed at the shoe and its interior is not filled with mud. This creates such buoyancy that the casing may have to be forced into the well. The buoyancy point of the casing string is given as follows:

$$P_a.g.h.(D^2ext - D^2int).\frac{\pi}{4} = \rho m.g.h.D^2ext.\frac{\pi}{4} \quad (3)$$

$$\rho a.(D^2ext - D^2int) = \rho m.D^2ext \quad (4)$$

With :

ρ_a = steel density (ppg) ;
ρ_m = density of drilling mud (ppg) ;
D_{ext} = outer diameter of casing (ft) ;
D_{int} = internal diameter of casing (ft).

Once the mechanical characteristics of casing have been calculated, their grade can be easily determined by reading directly from the driller's manual.

I.3.4 Cementing programme

Cementing an oil or gas well involves injecting a cementing fluid down the drill string or casing to a predefined section of the well annulus. The cementing fluid itself generally contains water, portland cement and various additives. The actual composition varies from one application to another. The functions of the various cementing jobs differ depending on the objectives and the technique applied (Wolfgang, 2016) :

- **Primary cementing:** This isolates the hydrocarbon formation from other formations and provides a firm seal and anchor for the wellhead equipment;
- **Cementing the liner:** This is used to cover an open section in the well;
- **Compression cementing:** This technique consists of forcing a pressurised cementing fluid into a confined area of the well in the annulus to correct defective primary cement work;
- **Cementing the plug:** This stops the production of water at the bottom of the well and adjusts the plug to provide a seat for the directional tools.

I.3.5 The drilling tool and downhole assembly programme

Drill selection is generally a fairly delicate task, but when it is carried out correctly, it has a major impact on the duration and total cost of the well.

I.3.5.1 Types of drill bits

There are several types of drill bit, each characterised by a particular shape and operating mechanism, and by a specific type of constituent material.

- **Tricone roller bits**

The cutting action of this bit can be described as follows: when the bit rotates at the bottom of the hole, the teeth are pressed against the formation and apply a force that exceeds the rock's resistance to compression (Azar, 2020). The advantages of rolling tricone bits over fixed bits are :

- They can withstand difficult drilling conditions;
- They are less expensive than fixed bits;
- They are more sensitive to pressure variations and therefore a better indicator of formation pressure.

- **Diamond bits**

They have no moving parts and can drill very long sections of hole when the right drilling conditions are established. Their drilling time in hard and abrasive formations is higher (Wolfgang, 2016). There are several types of diamond drill bit:

- **Polycrystalline diamond bits:** These are made from industrially manufactured thermostable diamonds which are mounted directly in the bit matrix.
- **Natural diamond bits:** These bits can drill through the hardest rock (highest compressive strength) but they drill more or less slowly and are very expensive. For this reason, they are used in very hard and abrasive formations that would destroy other types of bits over very short drilling distances (Baaziz, 2014).

- **Fishtail bits:** These types of bits are only applicable to soft formations, where they establish drilling progress by scraping the rock. Their advantages are that they require low pumping rates and are inexpensive.

- **Tri-blades:** These are most often used for very soft, sticky formations where tri-cone tools lose their effectiveness. It is important not to exert too much pressure on the tri-blades, otherwise you will end up with a "helical" borehole. In order to obtain cylindrical boreholes, it is advisable to carry out frequent sweeping operations.

- **Core bits:** Core bits, which have a ring-shaped structure with natural or industrial diamonds, are used for coring. Drilling with these tools leaves a column of rock in the middle of the bit, which is recovered by a device called a "corer".

The principle of classification and selection of the drilling tool is presented in appendices 1 and 2.

I.3.6 Well control and blowout prevention

During drilling operations, a pocket of pressurised fluid is sometimes intercepted, generally resulting in a flow of fluid from the formation into the well: this is known as a "kick". A blowout, on the other hand, is an uncontrolled flow of liquid or gas into the well. Fluid inflows can have several origins (Wolfgang, 2016) :

- Loss of circulation causing a drop in hydrostatic pressure;
- Drilling a horizon at abnormally high pressure with low-density mud;
- A reduction in hydrostatic pressure during swabbing;
- Insufficient drilling fluid in the well following the removal of the drill pipes.

To prevent a blowout from occurring, it is essential to detect fluid ingress as early as possible. Particularly in wells drilled in zones where the formation pressure is abnormally high, the various parameters for detecting fluid ingress must be observed continuously. These parameters are (Samba, 2018) :

- Increased pit volume;
- An increase in the return flow of drilling mud;
- Sludge returns even when the pumps are stopped;
- An increase in the chloride content of the sludge.

It should be noted that some of these parameters are sufficient on their own to identify a fluid inflow (for example, volume gain in the pit). When a fluid inflow is detected at the surface, a device called a Blowout Preventer (BOP) is used to stop the flow of fluid from the well. To cover all possible scenarios and control different fluid influx situations, different types of BOP are assembled at the wellhead to form a "BOP stack".

I.4 Directional drilling and deviation control

A well is said to be directional when it follows a predefined trajectory to cross specific targets. The principle is to deflect the drill bit in a controlled manner. The various methods used to steer the borehole can be classified as mechanical and hydraulic. The mechanical technique involves exerting an appropriate lateral force to deflect the tool; this method uses deflectors, downhole assemblies and downhole motors with a bending device. The hydraulic method, on the other hand, requires the use of a directed jet bit. Water or drilling mud is pumped through a jet oriented in the desired direction. The jet is the technique best suited to unconsolidated formations where compressive strength is relatively low (Farah, 2013).

I.4.1 Calculating the well trajectory

To start planning a directional well, we need to design the trajectory of the borehole to reach a given target. The first thing to do is to select the most economical design for the directional drilling profile.

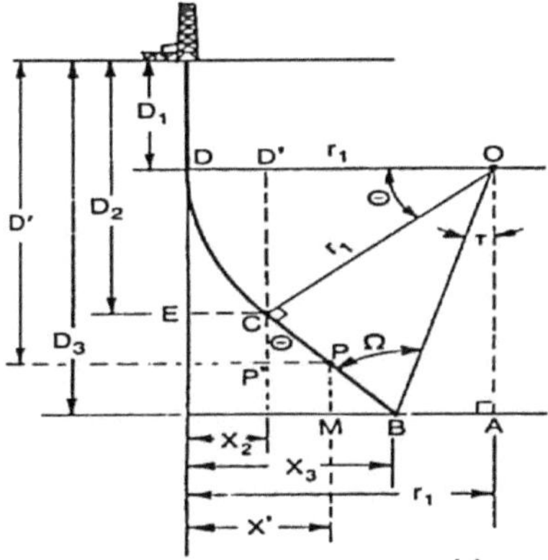

Figure 3 Geometry of a J-type directional drilling trajectory (Wolfgang, 2016)

The parameters that characterise a profile are determined using the following equations:

- Radius of curvature :

$$r_1 = \frac{180}{\pi}\left(\frac{1}{q}\right) \tag{5}$$

With :

q : rate of curvature inclination (°/ft) ;

- Maximum angle of inclination

$$\theta = \sin^{-1}\left(\frac{r_1}{\sqrt{(r_1 - X_3)^2 + (D_3 - D_1)^2}}\right)\tan^{-1}\left(\frac{r_1 - X_3}{D_3 - D_1}\right) \tag{6}$$

With :

D_1 kick off point depth (ft) ;
D_3 total depth of the first bend (build), (ft) ;
X_3 horizontal distance (ft) ;

r_1 radius of curvature (ft) , where $X_3 < r_1$;
θ Maximum angle of inclination (°).

- Depth and horizontal deviation along the curvature (Build-up) :

$$(D_i)_{\ Build} = D_1 + \frac{\theta_i}{q} \quad (7)$$

$$(X_i)_{\ Build} = r_1(1 - \cos\theta_i') \quad (8)$$

With $\theta_i'=\theta$ at the end of the curve.

$$\theta_i' = \sin^{-1}\left(\frac{D_i}{D_1 - r_1}\right) \quad (9)$$

With :

D$_i$: Vertical depth at point i along the curvature (build) and the straight section (hold) (ft).

- Depth and horizontal spacing of the straight section of the well :

$$(D_i)_{\ Hold} = D_1 + \frac{\theta}{q} + \frac{D_i - D_1 - r_1.\sin\theta}{\cos\theta}$$
(10)

$$(X_i)_{\ Hold} = r_1(1 - cos\theta) + (D_i - D_1 - r_1.sin\theta)tan\theta \quad (11)$$

With :

D_1 kick off point depth (ft) ;
D_3 total depth of the first bend (build), (ft) ;
X_3 horizontal distance (ft) ;
r_1 radius of curvature (ft) ;
θ Maximum angle of inclination (°).

Conclusion

The purpose of this chapter was to give a general introduction to the drilling system on a drilling rig and the elements involved in the design process for a drilling programme for a deviated well. At the end of this chapter, all the concepts essential to the design of our drilling programme are known and will serve as a basis in Chapter 2 for establishing the method for designing the drilling programme for the Doketi-1 well (A4-1).

Chapter II: MATERIALS AND METHODS

Introduction

In the previous chapter, we gave a general overview of the techniques used in the drilling industry to plan wells. Planning an oil well is a complex operation requiring multidisciplinary skills. The implementation of a drilling plan generally follows a method that varies according to the drilling objectives, the type of drilling, the information available and the context in which it is carried out. This chapter will focus on describing all the tools we have used to implement this work. We will also present in detail the method we adopted in this work in order to achieve each of the established objectives.

II.1 Equipment

The material used to carry out this work consists of a set of computer tools and documents:

- **The drilling manual (Schlumberger i-Handbook)**

This e-book enables you to quickly calculate and select the characteristics of the elements you need (calculating the volume of cement, selecting casings according to their API characteristics, etc.).

- **The CNPC (China National Petroleum Corporation) well proposal**

This is a document that is drawn up when there is a proposal to drill a well. This document contains the data needed to plan the drilling operation. It includes : The location of the well to be drilled, the type of well, its purpose, the characteristics of the formation to be drilled, etc.

- **A laptop computer**

This is our main working tool, and it was essential for us to obtain a high-performance machine (Toshiba Core i5, 8GB RAM, 1TB ROM) to ensure that we could make optimum use of the software's potential and write this report.

- **Microsoft Office Excel**

This is a spreadsheet program from the Microsoft Office suite. This software enabled us to organise our data and results in tabular form and perform calculations quickly and accurately.

- **MudWare software**

This is a drilling engineering software package developed by M-I SWACO. It is a collection of engineering programs related to drilling mud and drilling. This software contains most of the calculations generally used in the field when drilling and completing wells.

- **Halliburton E-RedBook**

This Halliburton software is used to perform a large number of calculations for drilling and completion operations, as well as conversions. It is an essential tool in operational planning situations, enabling rapid casing selection.

- **COMPASS software™**

This is oil industry software used to design deviated wells. In addition to defining the well trajectory, this software also enables anti-collision analysis to be carried out and the drill bit to be directed efficiently towards the target zones. In this way, the software can be used to design well pads efficiently, so as to limit the footprint of surface operations.

- **CasingSeat software™**

It is part of the Haliburton Landmark suite of software used to accurately determine casing installation depths. It allows rapid selection of casing beds based on pore pressure, formation fracture gradient and lithology.

- **Landmark StressCheck software™**

This is a casing design software package used to select optimum casings. The software features graphical design tools and algorithms that automatically generate minimum-cost solutions while guaranteeing optimum mechanical casing strength.

- **WellPlan software™**

This software provides the most comprehensive drilling engineering toolkit in the industry. Optimum use of the software reduces the cost of well design and planning, and enables you to anticipate risks and drill faster without compromising the operation.

- **WellCost software** ™

This software is designed specifically for drilling and completion and provides powerful tools for quickly developing drilling project estimates. The software automatically generates reports and graphs for easy analysis and presentation of results.

Figure 4 Data processing process and sequence of use of the various tools

II.2 Methodological approach

The workflow illustrated in Figure 5 covers all the steps required to develop a well plan. This work focuses on an API-defined well planning basis, which is primarily intended to provide standards for well design and operation in order to plan and construct the well safely.

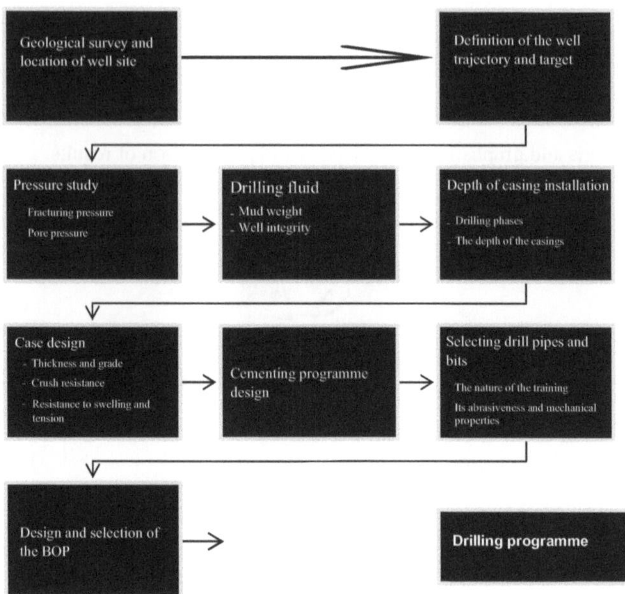

Figure 5 : Flux de travail pour la conception d'un programme de forage

II.2.1 Presentation of the study area

The Doba Basin is one of the major geological structures in Chad's great basin. It lies in the Logone Orientale region, in the Pendé département, between the 13ᵉ parallel north and the 19ᵉ meridian east, at coordinates 8°40' north and 16°51' east. Geophysical and geological studies carried out in this basin have revealed its stratigraphic structure and dated each of its constituent horizons. The work carried out by Louis (1970) shows that this basin is made up of several horizons, each of which was put in place at a very precise date in its history:

- 0 - 60 m: Quaternary sandy-clay horizon ;
- 60 - 120 m: Sandy horizon ;
- 120 - 200 m: Clay to sandy clay horizon ;
- 200 - 700 m: Sandstone horizon ;
- 700 - 1500 m: Cretaceous marl horizon ;
- 1500 - 4000 m: Cretaceous sandstone horizon ;
- 4,000 m: Base.

At the end of his work, Louis (1970) drew up a geological map of the Doba basin based on the data collected during his geophysical and geological reconnaissance campaigns. Figure 6 shows the geological map of the western part of the Doba basin (Louis, 1970).

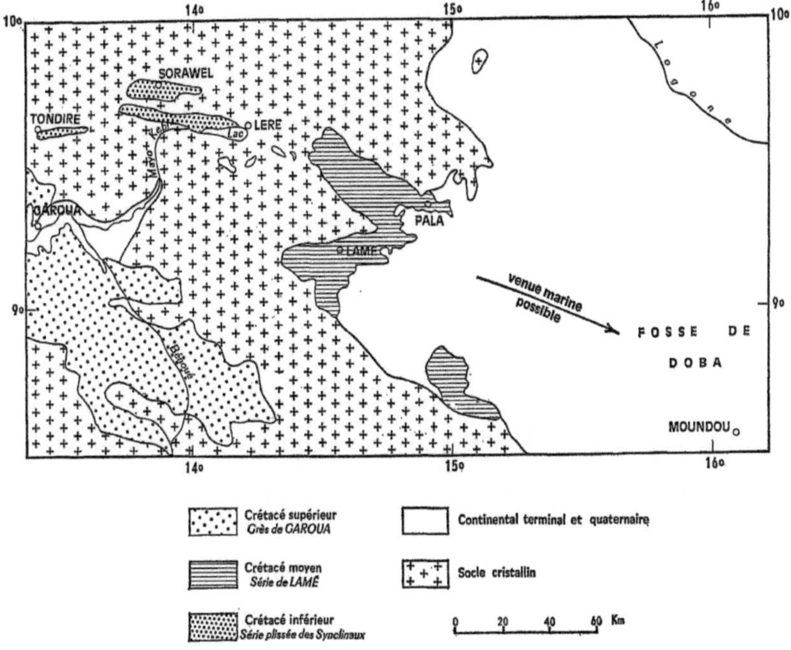

Figure 6 Geological map of the Doba basin (Louis, 1970)

A north-south geological profile of the Doba Basin (Figure 7) has been drawn up on the basis of the geological map of the basin and the information provided by the boreholes. This gives a better idea of the layout of the various strata in the basin and provides valuable information for directing future seismic exploration campaigns to identify potential hydrocarbon structures and traps. (Louis, 1970).

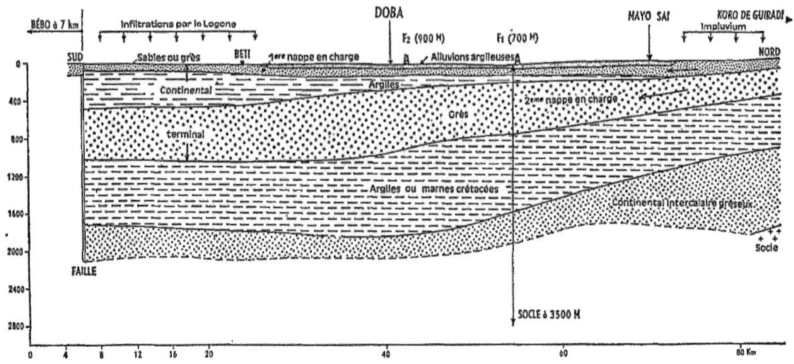

Figure 7 North-south geological profile of the Doba Basin (Louis, 1970)

II.2.2 Presentation of available data

The implementation of a drilling plan requires a minimum of geological, historical and technical data. This work was based on data supplied by CNPC (China National Petroleum Corporation) for the construction of a deviating well (Doketi-1) in the Doba basin for exploratory purposes. This data was provided in a "Well Proposal", a document that is drawn up when drilling is planned. The data and technical indications of the drilling programme are grouped into 4 in the Well Proposal:

- Basic design data ;
- Technical specifications and quality requirements ;
- Design engineering ;
- HSE indications.

The basic data used in this work is presented in Appendices 3 and 4.

II.2.3 Well trajectory design and target definition

In order to accurately reach the drilling target, it was essential to plan the drilling trajectory. The aim was to define an optimum trajectory for the Doketi-1 (A4-1) well that would intersect the predefined targets: Kedeni and Mangara at depths of 81,066.92 ft and 9,037.40 ft respectively. The type of trajectory selected to accurately reach our target is a "J" trajectory of the "Build and Hold" type. This type of trajectory is the most common in the industry and consists of the following parts: A vertical section, a Kick-off point, a Build-up section and a Hold section that extends to the target. The geometric parameters used to define it are essentially :

- The radius of curvature (R) ;
- The maximum angle of inclination of the oblique section (θ) ;
- The height and width of the curvature (D_2-D_1) and X_1 ;
- The height and width of the oblique section (D_3- D_2) and X_3.

In order to carry out this task efficiently, the data available in the Well Proposal and the Compass software™ were used. The input data used to define the drilling trajectory is shown in Table 1.

Table 1 Location of targets

Depth (ft)	Measured length of shaft (ft)	Training	Coordinates (m)	
			East	North
3592,51	3592,51	Kome Fm		
7204,92	7213,98	Doba	341669.43	997848.50
8106,69	8216,27	Kedeni Fm	341700.06	997690.68
9037,4	9384,12	Mangara Fm	341741.06	997479.71
9576,44	10078,12	Mangara 2	341802.57	997163.55
10054,79	10693,96	Mangara 3		

Once this data had been collected in an Excel workbook, it was entered into the Compass software™ . The output was graphs and diagrams illustrating the precise trajectory of the borehole. The table in Appendix 3 shows the results of the calculations based on a tilt rate of 3°/100ft. The workflow in this software can be summarised in 5 steps:

- Enter general information about the well and its location ;
- Define the position of the well in the location created ;
- Definition of the coordinates and depth of the various targets ;
- Input of directional drilling parameters such as inclination rate;
- Run the simulation and display the results.

Figure 8 shows the main display interface of the COMPASS software™ .

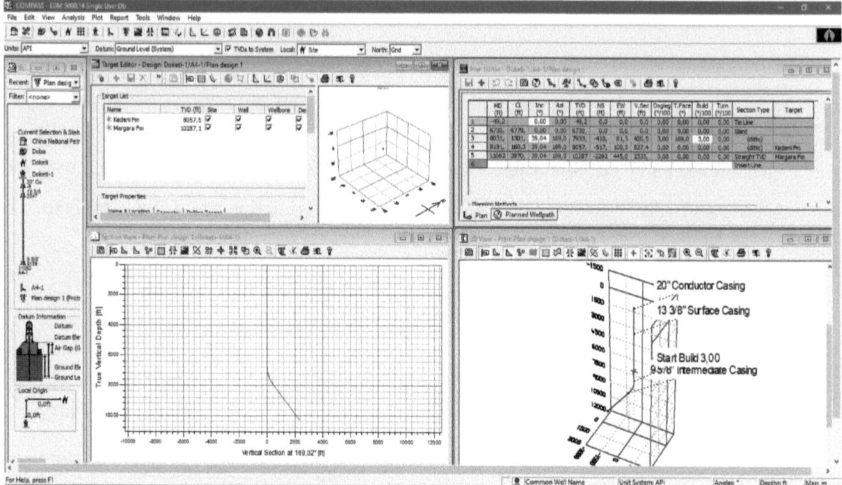

Figure 8 COMPASS software main display interface™

II.2.4 Well architecture and casing programme

The next phase of this work consists of casing sizing, selection and determination of casing shoe depths. In other words, in the steps that follow, we will estimate casing shoe depths based on the pressure in the formation and the nature of the terrain. The casing grades will also be determined by defining scenarios for the mechanical stresses to which the casing string may be subjected.

II.2.4.1 Determining hoof depth

Determining the depth of the shoes is a crucial stage in the casing programme. This depth is generally a function of the lithology, as unstable formations constitute high-risk zones for casing installation. However, given the pressure that can be encountered in the formations along the length of the well, taking this parameter into account would limit the risk of casing collapse and ensure the long-term integrity of the structure.

To achieve this objective, we chose CasingSeat™ as our design tool. This drilling engineering software is used to determine the optimum casing installation depth and to plan the drilling mud programme. The CasingSeat™ working method can be summarised in 6 steps:

- Enter general information about the location of the well ;
- Definition of the well trajectory ;
- Input of design parameters and definition of lithology ;
- Input of pore pressure, fracturing gradient and geothermal gradient ;
- Run the simulation and display the results.

Figure 9 shows the interface of the CasingSeat™ software used to enter the data and display the results.

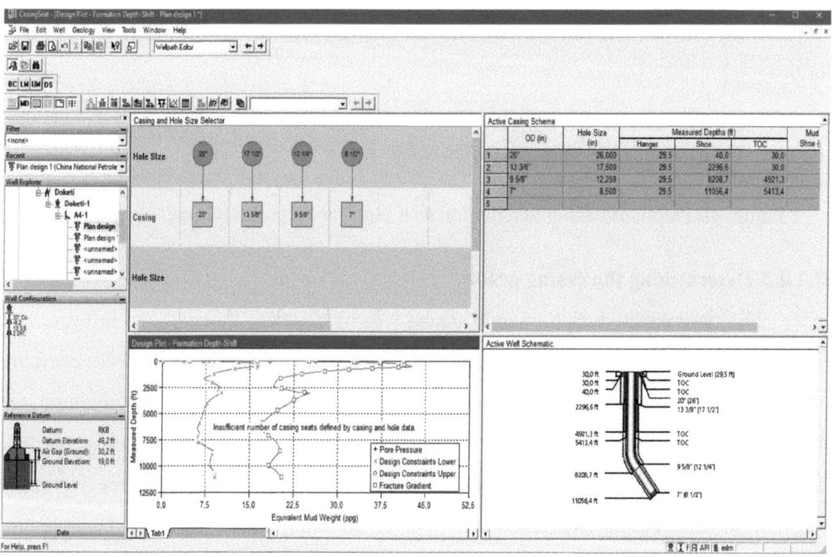

Figure 9 Main display window of the CasingSeat software™

The results provided by the CasingSeat™ software after running the simulation enabled us to define the casing installation depth as shown in Figure 10.

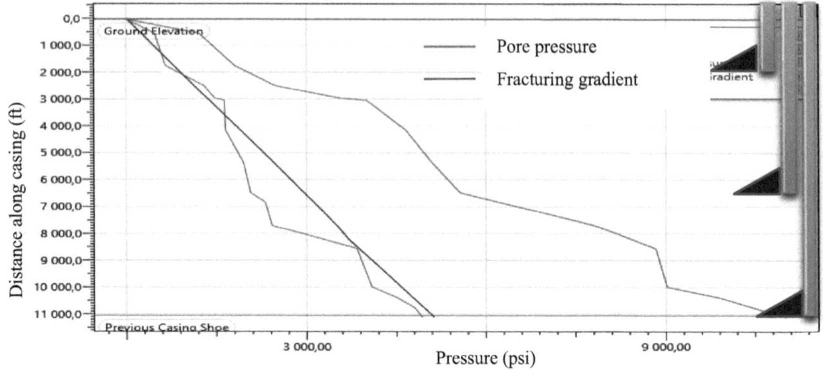

Figure 10 Depth of casing installation as a function of pore and fracture pressure

II.2.4.2 Determining the casing grade

The optimum selection of each casing was made using StressCheck software™ , taking into account the various stresses to which the casings may be subjected: crushing pressure, burst pressure and axial load. The following lines present the method for determining casing grades. Here we will consider the intermediate casing. This casing will have to be installed in a "problem zone" and will be subject to major stresses. Optimum selection of these casings will not only guarantee the safety and integrity of the structure, but will also considerably reduce the total cost of drilling. (Zenabou, 2018).

Selecting the grade of intermediate casing :

The grade of casing selected will depend on the stresses to which it will be subjected. Once these constraints have been determined, all that remains is to identify the corresponding grade in the Drilling Handbook by simple reading.

- **Burst pressure :**

$$Pb = Pf \, . F \quad (1)$$

P_b (psi): burst pressure ;

P_f = 4254.74 psi: maximum anticipated formation pressure for drilling the following section ;

F = 1,7 Safety factor ;

AN : P_b = 4254.74 × 1.7

$\boldsymbol{P_b}$**= 7 233.058 psi**

- **Crushing pressure :**

$$Pc = 0.052. \rho m. D. F \tag{2}$$

Pc (psi) : crushing pressure ;

F = 1.37 : Safety factor ;

ρ_m = 10.02 ppg: density of the drilling mud.

AN : P_c = 0,052 × 10.02 × 8208,7×1,37

$\boldsymbol{P_c}$**= 5,859.57 psi**

- **Axial load :**

$$Ft = 0,7854 \times (D^2ext - D^2int). \sigma. F \tag{4}$$

σ = 2209 psi : minimum casing pressure read in the drilling manual ;

D_{ext} = 10.625 in : external diameter of casing ;

D_{int} = 8.375 in: internal diameter of casing.

AN : F_t = 0,7854×(10,625 - 8,375) × 2209 × 2,98

$\boldsymbol{F_t}$ **= 221,023.78 lbf**

According to the drilling manual, grade N-80 casing has a crushing strength of 6,617.23 psi and a burst strength of 7,927.27 psi.

By applying the same method for each pipe phase, we can determine the grade of each casing. The values calculated for each casing and their grade are given in Table 2.

Name	MD (ft)	OD (in)	Burst (psi)	Crushing (psi)	Axial tension (lbf)	Gra de
Conductor tube	0 - 40,0	20"	1533	515,47	1076706	H-40
Surface tube	0 - 2296,6	13 3/8"	3763,04	2877,18048	1263938	K-55

Intermediate tube	0 - 8208,7	9 5/8"	7 233,058	5 859,57	221 023,78	N-80
Production leasing	0 - 11056,4	7"	7248,302	17199,6032	1485149	N-80

Table 2 Table of casing grades selected according to constraints

As mentioned above, the casing was designed using Landmark's StressCheck software. This software incorporates sophisticated design methods that enable reliable casing programmes to be designed at minimal cost. It also provides a variety of automated methods for specifying bursting stresses, crushing stresses and realistic axial loads, enabling optimisation of casing grade and cross-sections.

The workflow of the StressCheck software™ can be summed up in 8 steps, from data entry to displaying the results:

- Pore pressure input
- Fracturing pressure input ;
- Definition of the geothermal guard and surface temperature ;
- Definition of the various parameters contributing to bursting, crushing and axial casing load;
- Selection of casing characteristics ;
- Definition of the minimum cost of a K-55 steel grade casing ;
- Graphical adjustment of casing strength to different stresses;
- Display results and export reports.

These different parameters must be entered for each casing section, and the output is a report on the optimum selection of casing grade produced from the information entered.

Figure 11 shows the main display interface of the StressCheck software .™

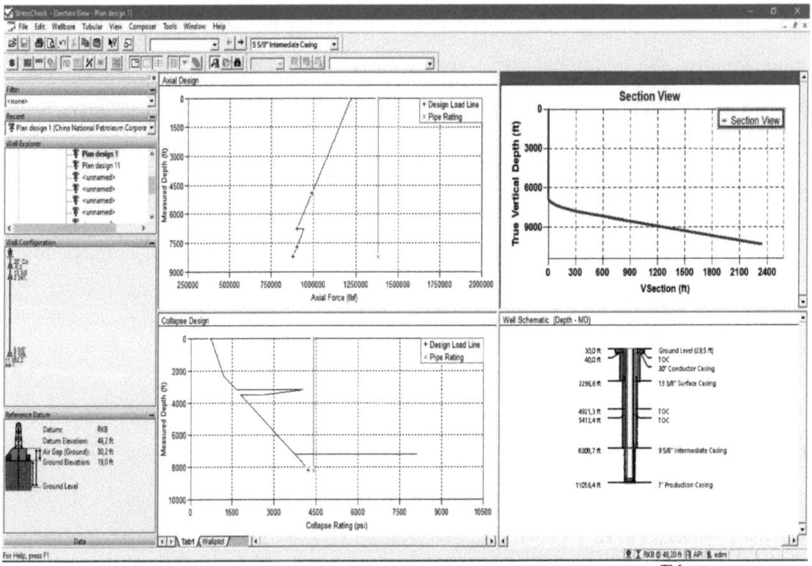

Figure 11 Main StressCheck software display window™

II.2.5 Cementing programme design

Like all the other phases in the construction of a well, cementing is a delicate stage that needs to be planned carefully to ensure the safety and integrity of the well. As prescribed by API, cement class is selected according to depth. Once the casing shoe depths are known, all that remains to be done is to determine the corresponding cement classes by reading the drilling manual. To minimise the risk of formation fracturing, loss of circulation or fluid ingress, the density of the cementing fluid must be the same as that of the drilling fluid during cementing operations (Wolfgang, 2016).

II.2.6 Drilling fluid programme design

The drilling mud programme is known to be a very critical phase in the well planning process. The density or weight of the drilling mud must be calculated taking into account the pore pressure and the fracture limit pressure of the formation. If the hydrostatic pressure of the mud is lower than the pore pressure, there will be kicking (fluid ingress), while if the hydrostatic pressure is higher than the formation's fracture limit pressure, there will be fluid loss and damage to the formation. In order to prevent these drilling problems from occurring, it is essential to first define a drilling mud window between the fracturing pressure and the pore pressure. To determine the optimum hydrostatic pressure for the drilling mud, we plotted the pore pressure and fracturing pressure as a function of depth. The area between the pore pressure curve and the fracturing pressure curve corresponds to the mud window. An additional curve corresponding to the variation in the hydrostatic pressure of the drilling mud as a function of depth is plotted. This last curve is closer to the pore pressure curve in order to reduce the risk of damaging the formation through fluid leakage (Enriquez, 2019).

Knowing the hydrostatic pressure required, we can determine the corresponding sludge density using equation 12 :

$$\rho_m = \frac{P_p}{0{,}052 \times Z} \tag{12}$$

With :

ρ_m Density of drilling mud ;

Pp Pore pressure ;

Z: Height of mud column (ft).

The graph in Figure 12 shows the drilling mud window determined from the hydrostatic pressure and fracturing pressure of the formation.

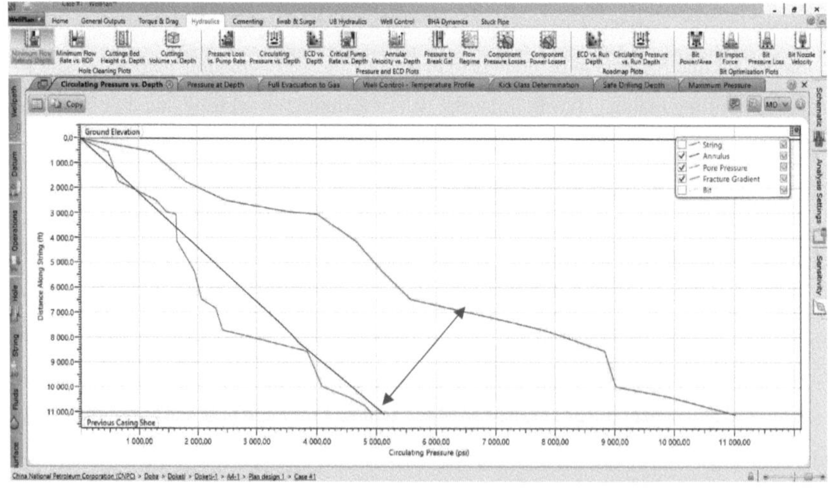

Figure 12 Graphic representation of the sludge window

II.2.7 Drilling tool programme design

The drill bit is one of the key elements in a drilling operation, as it not only determines the diameter of the hole, but also has a major impact on the speed at which drilling progresses. Optimum selection of this tool is therefore vital to ensure that drilling progresses quickly and efficiently. Its selection will depend mainly on the mechanical properties of the geological formations to be drilled (abrasiveness, shear strength, etc.).

In this study, the type of drill bit was selected on the basis of information about the geological nature of the formations through which the borehole was to pass (mainly marl, clay, sand and sandstone). The tool diameter was selected on the basis of the external diameter of predetermined casings (Appendix 2).

II.2.8 Well control programme design

The well control programme is the final phase in the planning of a drilling project, ensuring the safety of the drilling rig and the personnel on site. Well control equipment (BOP) is divided into five pressure classes:

2,000 psi, 3,000 psi, 5,000 psi, 10,000 psi and 15,000 psi. For the purposes of this work, the BOP operating pressure was determined on the basis of the maximum pressure likely to be encountered in the well during drilling (information available in the Well Proposal). This selection was made taking into consideration the standard classification "Well Control Regulation of Oil and Gas Drilling (CNPC)" which is a standard recommended by the State of Chad. As stated in this standard, the operating pressure of the BOP must be at least 10% higher than the maximum pressure likely to be encountered during drilling. Since the maximum pressure likely to be encountered is known, we determined the BOP operating pressure by adding 10% of this pressure, i.e. a total of 5,280 psi.

II.2.9 Evaluation of drilling time

Estimating drilling and completion time is a variable that is governed by different activities during drilling. An accurate estimate of drilling time is necessary for the preparation of an expenditure authorisation (AFE). The total drilling time is estimated on the basis of several factors: the drilling time, the time taken to assemble/dismantle the drill string, the number of bits and the connection time.

- **Estimated drilling time**

Drilling time is a function of the rate of penetration (ROP) of the drill bit into the formation. For a given formation, the ROP is inversely proportional to the rock's compressive and shear strength. In addition, rock strength tends to increase with burial depth (Hossain, 2015) .

- **Estimated assembly/disassembly time for drill string**

When estimating well drilling time, a second major component of the time required to drill a well is the time taken to assemble and disassemble the drill string; this is the time required to change a drill bit and resume drilling operations. The time required for this operation depends mainly on the depth of the well, the type of rig and the drilling practices adopted. This time can be approximated as follows (Hossain, 2015) :

$$t_t = 2\left(\frac{t_s}{L_s}\right) D_t \tag{13}$$

With :

t_t Connection/disconnection time (hr) ;

t_s Time required to manoeuvre a drill pipe stand (hr) ;

L_s Length of rod stand (ft) ;

D_t Depth of well.

For the purposes of this work, drilling time was estimated using WellCost^{TM} software. This is a software package designed specifically for estimating the time and cost of drilling and completion operations. The WellCost software^{TM} provides powerful tools for producing drilling project estimates quickly and easily. The procedure for using this software to estimate drilling time can be summarised in 4 steps:

- Enter general information about the project ;
- Definition of the well trajectory, lithology and casing plan ;
- Setting probabilistic calculation parameters (Monte Carlo method) ;
- Definition of the phases and activities of the drilling programme and their duration.

Figure 13 shows the interface of the WellCost^{TM} software, from which the phases, activities and their respective durations have been defined.

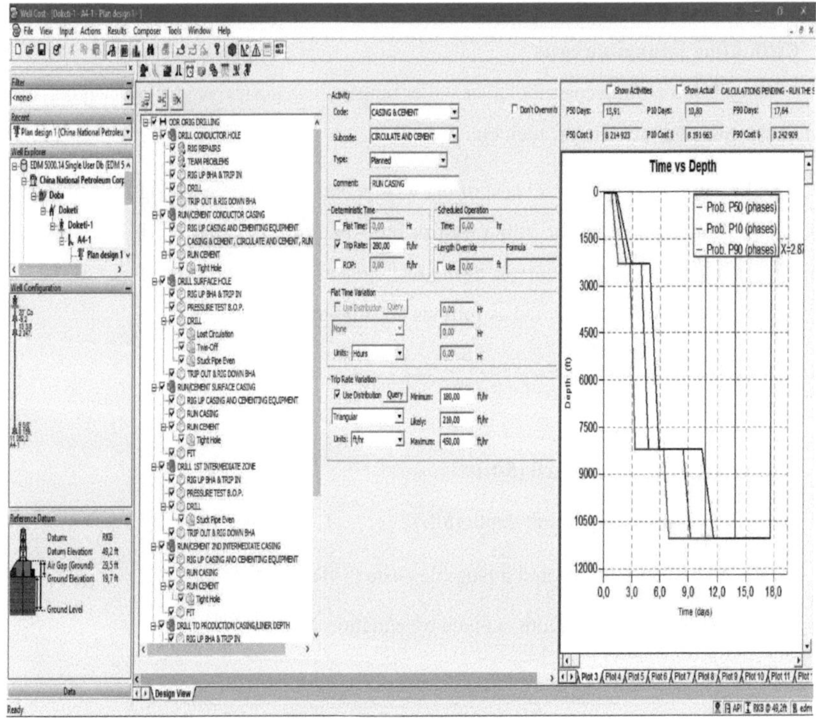

Figure 13 WellCost software interface™

II.2.10 Estimating the cost of drilling and drawing up a budget

An accurate estimate of the cost of drilling is essential both for budgeting purposes and for drawing up a more detailed authorisation to spend (AFE). Companies must therefore make the best possible use of the funds available. Underestimating project costs could lead to budget shortfalls and the risk of delays, while overestimating could result in the misallocation of funds that could have been used on other projects.

For the purposes of this work, total drilling costs include all the elements required to carry out the drilling programme. However, well drilling costs can be subdivided into 2 classes, each subject to different tax rules. Tangible costs and intangible costs (Kitchel et al., 1997).

- **Estimating tangible costs :**

Tangible drilling costs generally include the direct cost of drilling equipment. They refer to the products used on the well. These costs include : The cost of the casing, the cost of the wellhead...

- **Estimating intangible costs :**

Intangible drilling costs include a wider range of categories: payroll, surveying, tools and materials, contract labour, fuel, etc.

The estimate of the actual cost of the well is obtained by integrating the drilling and completion times provided for in the well design. The time required to drill a well has a significant impact on many elements of the well cost estimate. The overall cost of the well excluding production is calculated as follows (Wolfgang, 2016) :

$$C_{owc} = C_f + C_o \quad (14)$$

With :

C_{owc} Overall cost of well ($/ft) ;

C_f Drilling cost per unit depth ($/ft) ;

C_o All other tangible and intangible costs ($/ft).

The cost of drilling per unit depth is given by equation 15 :

$$C_f = \frac{C_b + C_r(t_d + t_c + t_t)}{\Delta D} \quad (15)$$

With :

C_b Cost of drill bit ($) ;

C_r Platform operating cost per unit of time ($/hr) ;

t_d Bit rotation time (hr) ;

t_cStop time (hr) ;

t_t : Drill string assembly/disassembly time (hr) ;

ΔD: Interval of the formation drilled (ft).

The above equation has several assumptions:

- It ignores the risk factors associated with drilling operations, the rate of inflation and the costs of environmental effects;
- The results of the cost analysis sometimes need to be tempered by technical judgement.

As part of this work, the WellCost™ software package was used to estimate the duration and cost of drilling. This software was chosen because of its calculation method, which incorporates the Monte-Carlo method. This method refers to a family of algorithmic methods designed to calculate an approximate numerical value using random processes, i.e. probabilistic techniques. The workflow on this software for estimating the cost of drilling can be summarised in 6 steps:

- Enter general information about the project ;
- Definition of the well trajectory, lithology and casing plan ;
- Setting probabilistic calculation parameters (Monte Carlo method) ;
- Definition of the phases and activities of the drilling programme and their duration ;
- Definition of the costs associated with each element of the drilling programme ;
- Simulation and display of results.

Figure 14 shows the interface of the WellCost™ software used to define the cost of each element of the drilling programme.

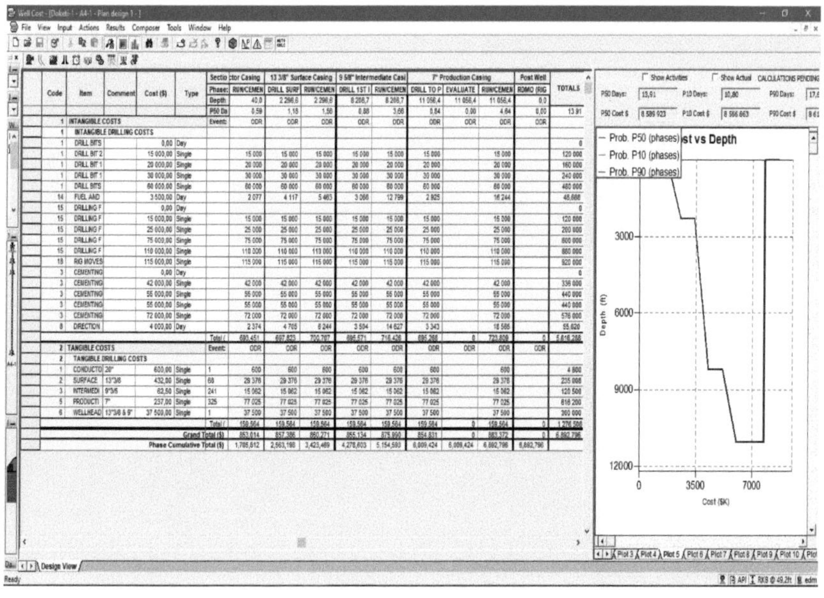

Figure 14 Drilling programme component cost entry window

Conclusion

This chapter presents the material used in this work and the method used to achieve the initial objectives. It has been shown that well planning is a complex operation that requires in-depth knowledge of drilling engineering. The development of a reliable drilling plan and the estimation of its cost necessarily involve the use of computer tools, in particular specialised oil industry software incorporating historical data. Efficient use of the potential of this software would enable oil industry operators to save time and resources while achieving excellent accuracy in the planning of their projects.

Chapter III: RESULTS AND INTERPRETATION

Introduction

The application of the methods and techniques described in the previous chapter has made it possible to draw up a detailed drilling programme for the Doketi-1 well and to estimate its duration and cost. This chapter will therefore present the results obtained from both the technical and economic points of view, and provide an interpretation and commentary on each result obtained.

III.1 Rig selection and main drilling system equipment

In this work, three major elements of the drilling system were dimensioned: the lifting system, the circulation system and the rotation system.

- **The lifting winch**

The role of the winch is to provide the lifting and braking power required to raise and lower the drill string and casing column. For the purposes of this work, the power required from the winch was determined on the basis of a speed of the moving block V_b = 590 ft/min, a weight on the hook W= 72,186.36 lb, and a system power efficiency E=1.

$$P_h = \frac{W.V_b}{33000.E} \tag{16}$$

P_h = 72186,3581×590/(33000×1)

$\boldsymbol{P_h}$ **= 1290.6 HP**

- **The sludge pump**

Two types of mud pump can be distinguished today: the duplex pump and the triplex pump, both equipped with reciprocating pistons. Triplex pumps are generally lighter, more compact, easier to maintain and less expensive than duplex pumps; these were the main reasons why this type of pump was selected for this drilling programme. Determining the pump power required was therefore essential to ensure efficient circulation and transport of the mud and cuttings. This was determined by assuming a discharge pressure Δp = 900 psi, and a pumping rate q = 2,100 gpm.

$$P_h = \frac{\Delta p.q}{1714} \quad (17)$$

$P_h = 900 \times 2100/1417$

$\boldsymbol{P_h} = \mathbf{1\ 102.7\ HP}$

- **The rotation system**

The function of the rotary system is to transmit rotation to the drill string and consequently to rotate the drill bit. Several parameters are used to characterise this system, in particular the rotation power, which will determine the equipment's ability to rotate the drill string efficiently. The power of the rotation system was determined by taking into account the rotation speed N = 200 rpm and the maximum torsion of the drill string T = 38,243.34 ft-lb (Wolfgang, 2016).

$$P_r = \frac{T.N}{5250} \quad (18)$$

$P_r = 38\ 243{,}335 \times 200/5250$

$\boldsymbol{P_r} = \mathbf{1{,}456.89\ HP}$

Table 3 Summary of rig selection results

Features	Value
Type	Onshore drilling platform
Winch power	1,290.6 HP
Rotation power	1,456.89 HP
Sludge pump power	Ph = 1 102.7 HP
Derrick height	45,5 m
Maximum load on the hook	3 150 KN

III.2 Well architecture and trajectory

The path followed by a well is influenced by a number of factors, including the presence of obstacles and the drilling objectives. The trajectory of the Doketi-1 well was designed to intersect the different targets indicated in the Well Proposal, respectively the Kedeni geological formations at a depth of 8,216.27 ft (Target 1) and Mangara at a depth of 9,384.12 ft (Target 2). The well's profile follows an oblique "Build and Hold" trajectory, which allows it to intersect a larger portion of the target zones compared with vertical wells.

The trajectory of the Doketi-1 well designed using COMPASS software™ is shown in Figure 15.

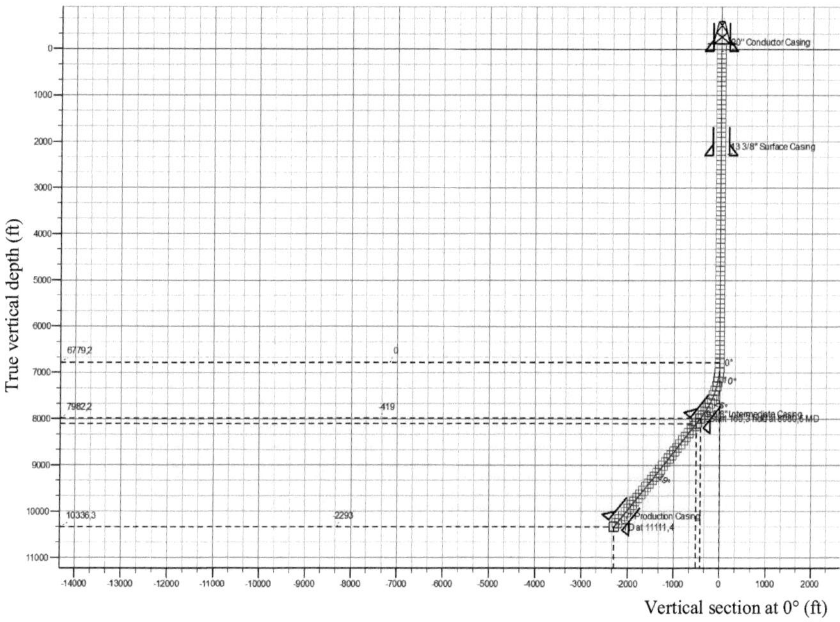

Figure 15 Trajectory of the Doketi-1 well (A4-1)

The "J" type well profile shown in Figure 15 consists of 3 sections: A vertical section to a depth of 6,779.2ft; a curved section (Build) with a 3°/100ft inclination rate that extends to a depth of 7,982.2ft; and a sloping section (Hold) that extends to the final depth of 10,336.3ft.

Figure 16 shows the architecture of the Doketi-1 well designed using Halliburton e-RedBook software™ . The Doketi-1 well consists of 4 sections: a 20" conductor pipe, a 13 3/8" surface casing, a 9 5/8" intermediate casing and a 7" production casing.

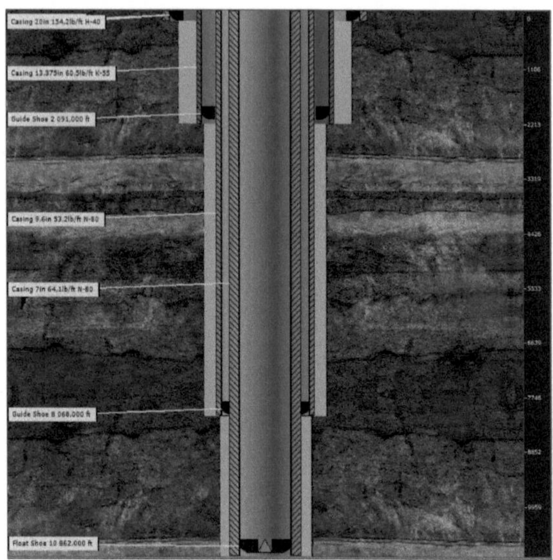

Figure 16 Architecture of Doketi-1

III.3 Casing programme and well architecture

The casing programme was designed using CasingSeat™ to determine shoe depth and StressCheck™ to determine casing grade. The aim was to design a reliable casing programme at minimum cost that would guarantee the long-term integrity of the well.

III.3.1 Hoof depth

In developing this casing programme, the depth of the shoes for each casing section was determined by taking into account the lithology and by analysing the fracturing gradient and pore pressure curves generated by the CasingSeat™ software. An analysis of these curves reveals the existence of sudden increases in the fracturing gradient at certain depths; these depths correspond to optimum casing installation zones. Figure 25 shows the casing shoe depths.

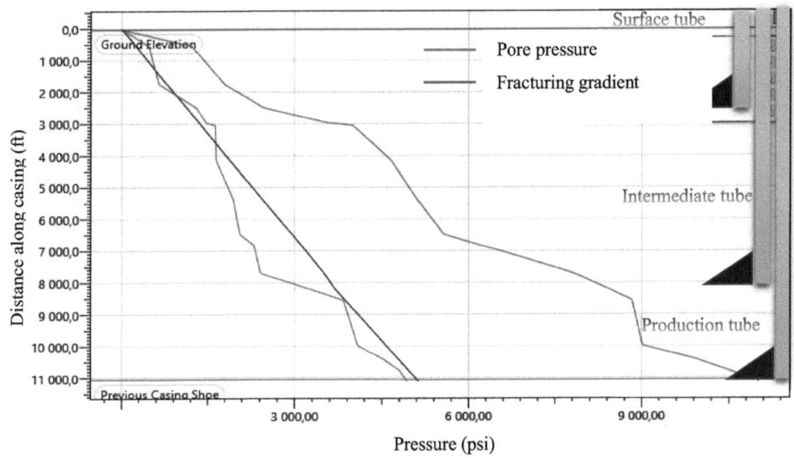

Figure 17 Depth of casing installation

The shoe depths shown in figure 16 are the result of an analysis of the formation's fracturing gradient curve. The setting depth for the surface casing is 2,296.6 ft, that for the intermediate section is 8,208.7 ft and 11,056.4 ft for the production casing.

III.3.2 Selection of casing diameter

The casing diameter was selected on the basis of the API standards available in the Drilling Handbook and using the CasingSeat software^{TM} . The selection was made starting with the smallest to the largest diameter sections. The result of the casing diameter selection for each section is shown in Table 4 :

Table 4 Dimensions of casings

Section	Length (ft)	Diameter (in)
Driver leasing	0 - 213,36	20"
Surface leasing	0 - 2296,6	13 3/8"
Intermediate leasing	0 - 8208,7	9 5/8"
Production leasing	0 - 11056,4	7"

La détermination de la profondeur des sabots et du diamètre de casing nous a permis de concevoir l'architecture du puits Doketi-1 telle que présentée sur la figure 17.

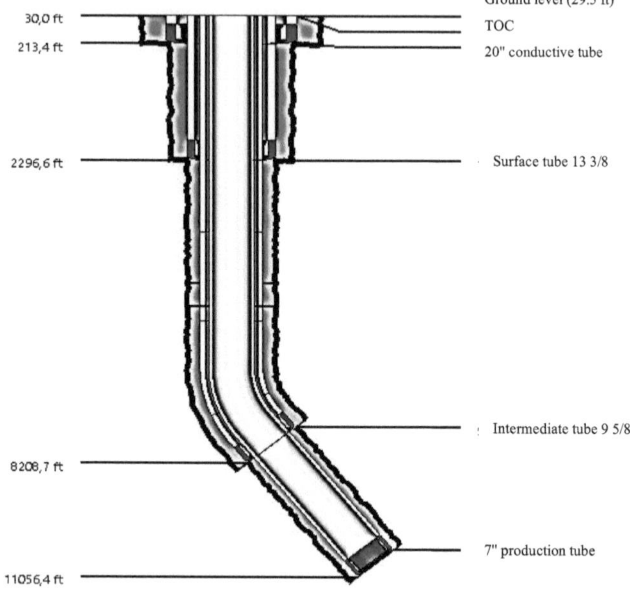

Figure 18 Architecture of the Doketi-1 (A4-1) well

III.3.3 Selection of the casing grade

As mentioned above, determining the grade of casing requires an in-depth study of the mechanical stresses to which they will be subjected. The main factors taken into consideration in this study are: burst pressure, crushing pressure and axial load due to the weight of the casing column. An interpretation of the stress curves for the intermediate casing will be made, given that the analysis method is the same for all casing sections.

Selection of the intermediate casing grade :

This section, also known as the "problem zone", is the site of phenomena that are generally the cause of many problems during drilling. Selecting the right casing for this section is therefore a delicate task.

- **Burst pressure**

These are the stresses exerted on the internal walls of casings by the fluids they contain. Figure 18 shows the variation in burst pressure as a function of depth.

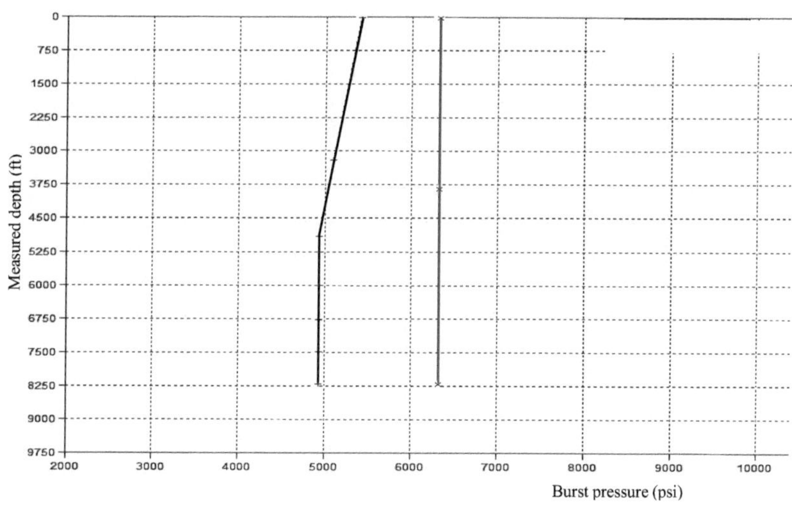

Figure 19 Burst pressure with a safety factor of 1, 7

Figure 18 shows that the burst pressure is highest in the upper part of the casing string; this is due to the fact that the formation pressure is generally low at the surface. Determining this constraint assumes a scenario where the formation pressure is at a maximum (a fluid inflow, for example); in this way, the value obtained will be such that the casing can withstand the maximum burst pressure. Figure 18 shows a burst pressure of around 7,927 psi (red curve) corresponding to a steel grade of L-80.

- **Crushing pressure**

The crushing pressure corresponds to the stresses exerted on the external walls of the casing column by the fluids present in the annular space; it is therefore maximum at the casing shoe and almost zero towards the surface. To determine it, we assume that the casing string is almost empty (loss of fluid in the formation); in this way, the value obtained will be such that the casing can withstand the maximum crushing pressure.

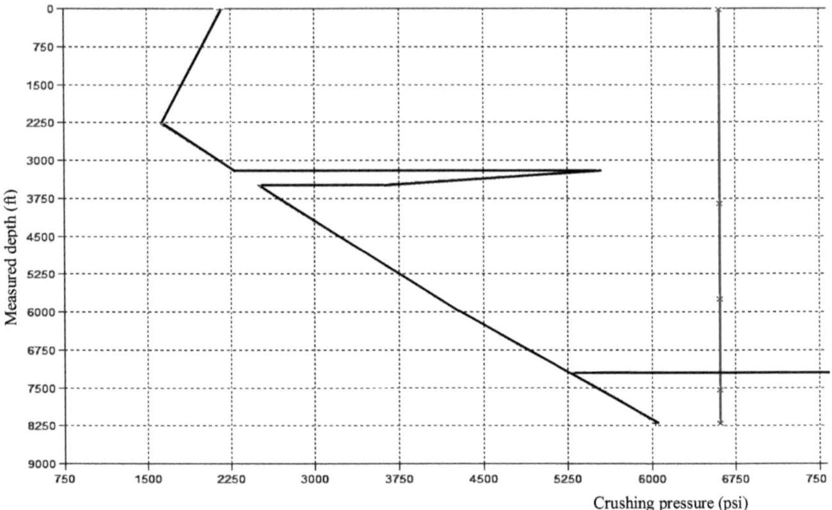

Figure 20 Crushing pressure with a safety factor of 1. 37

Figure 19 shows a crushing pressure of approximately 6,617 psi (red curve). Corresponding to steel grade L-80.

- **Axial load**

This is the tensile stress due to the weight of the casing string itself. This stress is generally counterbalanced by buoyancy, but is sometimes significant.

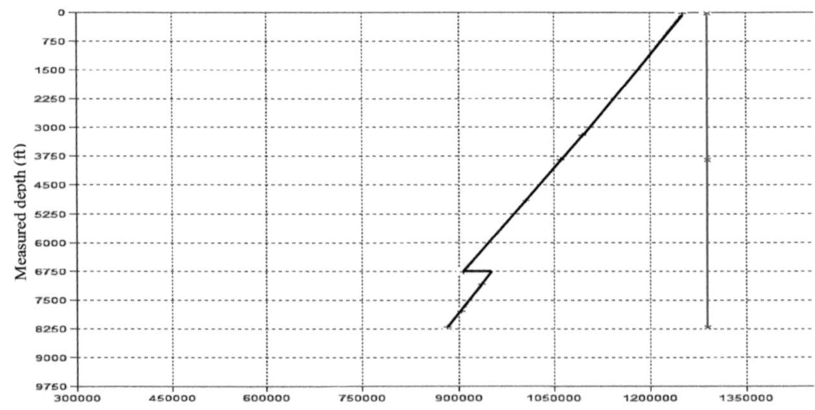

Figure 21 Axial load with a safety factor of 1. 3

Figure 20 shows an axial load of approximately 1,350,000 psi (red curve). Corresponding to steel grade N-80.

Table 5 summarises the mechanical characteristics of each casing section.

Table 5 Summary of casing section grades according to stresses

Section	MD (ft)	OD (in)	Burst (psi)	Crushing (psi)	Axial tension (lbf)	Grade
Conductor	0 - 213,36	20"	1 533	515,47	1076706	H-40
Surface	0 - 2296,6	13 3/8"	4082,24	2 101,75	1263938	K-55
Intermediate	0 - 8208,7	9 5/8"	7 927	6 617	1004719	N-80
Production	0 - 11056,4	7"	18 437,50	19 295,41	1485149	N-80

III.4 Cement selection

The cementing technique used is primary cementing. As mentioned in Chapter 1, its purpose is to isolate the formations, protect and secure the casing in the well, prevent collapse of the well walls, provide a firm seal and anchor for the wellhead equipment and protect the casing from corrosion by sulphate-rich formation water. The density of the cementing fluid was set so as not to exceed the fracturing pressure of the formation. The cementing programme for the Doketi-1 well was carried out in 3 phases, each using cementing fluids with different properties.

Table 6 Summary of cement selection

Phase	Class	Mass Vol. (g/cm)^3	Traffic speed (m/ft)	Additives	Sulphate resistance
Surface	G	1,85	1 à 1,3	Strengthening 5%.	Average
Intermediate	G	1,60	1 à 1,3	Dispersant 0.2	Average
Production	G	1,50	1 à 1,3	1.2% desiccant	High

III.5 Selection of drilling tool and downhole assembly

Each phase of the drilling programme corresponds to a downhole assembly, the characteristics of which depend on the lithology, the diameter of the section of the well to be drilled, the drilling trajectory, etc. The drilling programme for the Doketi-1 well is therefore subdivided into 3 phases, each corresponding to a given hole diameter and downhole assembly.

Section	MD (ft)	Diameter (in)	Type	Bottom hole assembly
Surface	2296,6	17 1/2"	TMT	Drill bit, Motor, Stabiliser, UBHO, DC, HWDP, DP
Intermediate	8208,7	12 1/4"	PDC	Drill bit, Motor, Stabiliser, UBHO, NMDC, DC, HWDP, DP
Production	11056,4	8 1/2"	PDC	Bit, Motor, Float, UBHO, NMDC, DC, HWDP, DP

Table 7 Tool selection and downhole assembly

III.6 Drilling mud programme

The drilling mud programme for the Doketi-1 well was designed taking into account the nature of the formations through which the well will pass, the pore pressure and the fracturing gradient of the formation. As each section of the well is characterised by a given lithology and formation pressure, we divided the mud programme into 4 phases:

- **The cross-sectional area of the conductive tube** :

As this section is installed by ramming, it does not require the use of mud, so fresh water circulation is an effective solution for cleaning the hole.

- **The surface section :**

For this section, a water-based slurry with a density ranging from 1.03 to 1.08 g/cm^3 was chosen. This weight of slurry, which falls within the slurry window determined above, will exert sufficient pressure to counterbalance the pore pressure without fracturing the formation.

- **The intermediate section :**

The intermediate section should be drilled using a denser water-based slurry with a density of between 1.08 and 1.2 g/ml.cm^3.

- **The production section :**

Because of the high pressure in this section, it will have to be drilled with a denser mud than that used for the previous two phases, i.e. a density of between 1.2 and 1.3 g/cm^3 . This mud will produce sufficient hydrostatic pressure to maintain the fluids in the formation without fracturing the rock. Figure 21 shows the mud window from which the drilling mud programme was developed.

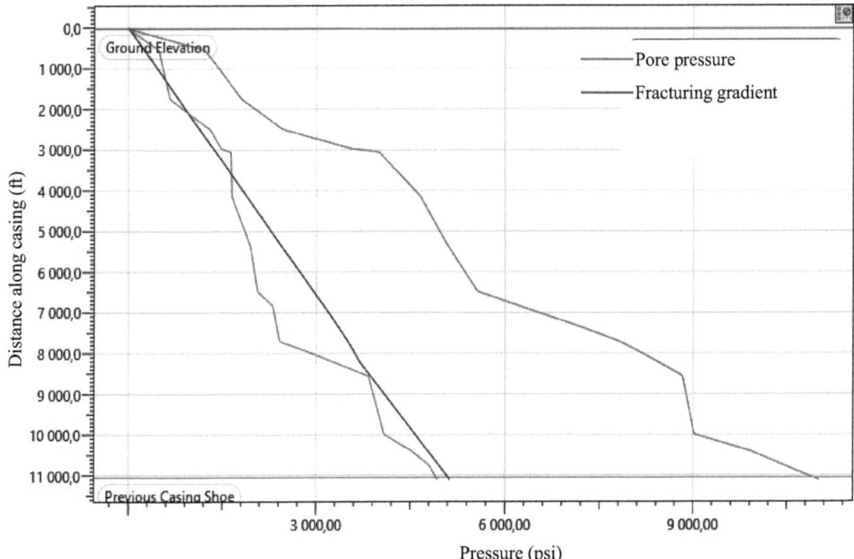

Figure 22 Drilling mud window

However, it should be noted that the properties of formations and the fluids they contain vary with depth, which is why a number of chemical additives need to be added to the drilling fluid to optimise its effectiveness and minimise drilling problems linked to mud incompatibility. Table 8 summarises the drilling mud programme:

Table 8 Mud programme for the Doketi-1 (A4-1) well

Section	Depth (ft)	Type	Mass Vol. (g/cm^3)	pH	Viscosity (cp)	Yield (lbs/100ft^2)
Conductor	213,36	Fresh water	1	7,5	1	————
Surface	2296,6	Water	1,03 - 1,08	8 - 9	15 - 30	10 - 30

Intermediate	8208,7	Water	1,08 - 1,2	8 - 9	15 - 30	10 - 30
Production	11056,4	Water	1,2 - 1,3	8,5 - 10	15 - 30	10 - 30

III.7 Well control programme

The aim of this programme is to size the main well control equipment capable of containing the fluid pressure in the well in the event of a kick. This equipment is selected on the basis of the maximum pressure likely to be encountered in the well. The main piece of equipment to be sized here is the BOP operating pressure. This corresponds to the maximum pressure that the BOP is capable of containing in the well. Based on the "Well Control Regulation of Oil and Gas Drilling" (CNPC) standards, the BOP operating pressure must be at least 10% higher than the maximum pressure likely to be encountered in the well. Given the maximum formation pressure (4,800 psi), the BOP operating pressure was estimated at 5,280 psi.

III.8 Estimated drilling time

The duration of drilling the Doketi-1 well was estimated using the WellCost software™ using the Monte Carlo method (probabilistic method). With the phases, activities and risks that could lead to lost time defined in advance, the software enabled us to generate a duration for each phase, each corresponding to a given degree of probability.

Table 9 shows the phases and activities that make up the drilling programme:

Table 9 Phases and activities of the drilling programme

Phase	Activity	Depth (ft)		Deterministic Time		Prob. time	
		From	À	Hr	Cum. days	Hr	Cum. days
DRILLING THE SURFACE SECTION	Staff-related problems	0,0	0,0	0,00	0,00	60,64	5,80
	Threshing the conductor	29,5	29,5	24,00	1,00	24,54	6,82
	Descending the garnit.	40,0	40,0	1,57	1,07	1,56	6,89
	Threshing the conductor	40,0	40,0	3,60	1,22	3,63	7,04
	Loss of circulation	0,0	0,0	0,00	4,35	0,51	9,35
	Stem distortion	0,0	0,0	0,00	4,35	0,30	9,37
	Stem jamming	0,0	0,0	0,00	4,35	1,44	9,43
	Remounté de la garnit.	2 296,6	2 296,6	0,00	4,35	0,00	9,43
CASING AND CEMENTING OF THE SURFACE SECTION	Preparation of casing and cementing equipment	2 296,6	2 296,6	1,57	4,41	1,57	9,49
	Pipework	2 296,6	2 296,6	24,00	5,41	22,91	10,45
	Cementing	2 296,6	2 296,6	24,00	6,41	24,48	11,47
	Cement setting	0,0	0,0	0,00	6,41	1,50	11,53
	BOP installation and pressure test	2 296,6	2 296,6	24,00	7,41	24,05	12,53
DRILLING THE INTERMEDIATE SECTION	Descending the garnit.	2 296,6	2 296,6	1,57	7,48	1,57	12,60
	Loss of circulation	0,0	0,0	0,00	17,17	0,56	19,36
	Stem distortion	0,0	0,0	0,00	17,17	0,35	19,38
	Stem jamming	0,0	0,0	0,00	17,17	1,61	19,45
	Remounté de la garnit.	8 208,7	8 208,7	0,00	17,32	0,00	19,60
CASING AND CEMENTING OF THE INTERMEDIATE SECTION	Preparation of casing and cementing equipment	8 208,7	8 208,7	1,57	17,39	1,58	19,66
	Pipework	8 208,7	8 208,7	72,00	20,39	71,01	22,62

	Cementing	8 208,7	8 208,7	48,00	22,39	48,16	24,63
	Cement setting	0,0	0,0	0,00	22,39	1,49	24,69
	BOP installation and pressure test	8 208,7	8 208,7	24,00	23,39	23,99	25,69
DRILLING IN THE PRODUCTION SECTION	Descending the garnit.	8 208,7	8 208,7	1,57	23,45	1,58	25,76
	Loss of circulation	0,0	0,0	0,00	23,45	0,60	28,21
	Cement setting	0,0	0,0	0,00	23,45	0,31	28,22
	Stem jamming	0,0	0,0	0,00	23,45	1,52	28,29
	Remounté de la garnit.	11 056,4	11 056,4	0,00	23,45	0,00	28,29

Drilling of the Doketi-1 well will take place in 3 main phases: the surface phase, the intermediate phase and the production phase. The flowchart shown in Figure 22 describes the sequence of the various phases and activities that make up the drilling programme for the Doketi-1 well.

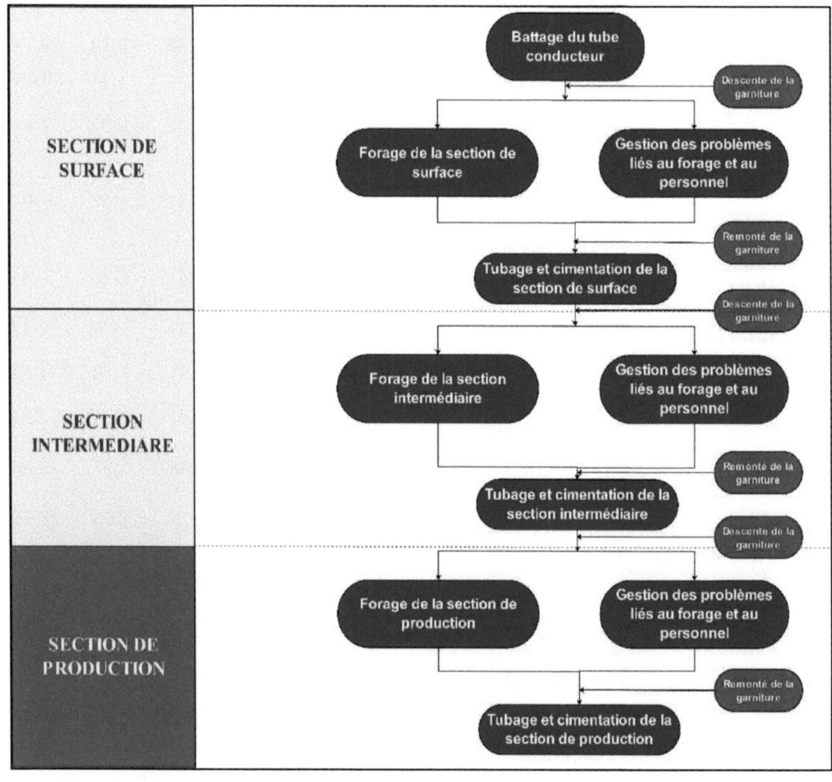

Figure 23 Flow chart of phases and activities

The graph in Figure 22 shows the vertical progression of the borehole as a function of time.

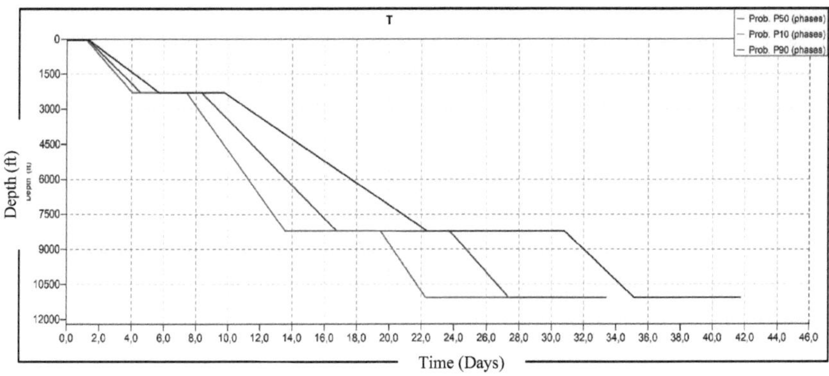

Figure 24 Vertical progress of drilling as a function of time

This graph describes the vertical evolution of the borehole as a function of time. The Monte Carlo probabilistic method was used to generate 3 curves, each corresponding to a given degree of probability (10%, 50% and 90%). The appearance of the curves in staircase form reveals four depth levels corresponding to the four phases of drilling. The first phase, in which the conductor pipe is rammed, lasts 5 days. The second drilling phase for the surface section lasts 7 days, the intermediate section lasts 12 days, and the production section lasts 8 days. Assuming a 50% probability level, the total drilling time is estimated at around 33.39 days.

III.9 Estimating the cost of drilling and drawing up a budget

Assessing the cost of drilling is a crucial stage in the drilling programme. It helps to determine the requirements for each phase of drilling, to establish the budget and to plan expenditure effectively during the execution of the programme. A good estimate of the cost of drilling will therefore enable better management of the company's resources.

As part of this work, the cost of drilling was estimated using the WellCost software™. This software incorporates probabilistic calculation methods to estimate the cost of the project based on the information entered. In its default configuration, WellCost™ divides the needs of the drilling programme into 2 groups: tangible costs and intangible costs.

- **Tangible costs**

The tangible costs of drilling correspond to the equipment such as the wellhead, casings, etc. used directly on the well. The tangible costs of the drilling programme for the Doketi-1 well are made up of :

- Driving the conductive tube, which costs $4,800 (less than 1%);
- Surface casing, $235,008 (18%) ;
- Intermediate casing, $120,500 (9%);
- Production casing, $616,200 (48%) ;
- The wellhead, $300,000 (24%).

Thus, assuming a 50% probability, the cumulative value of tangible costs amounts to $1,276,510, or 21% of the total cost of drilling. Figure 23 shows a breakdown of the tangible costs and the corresponding percentages.

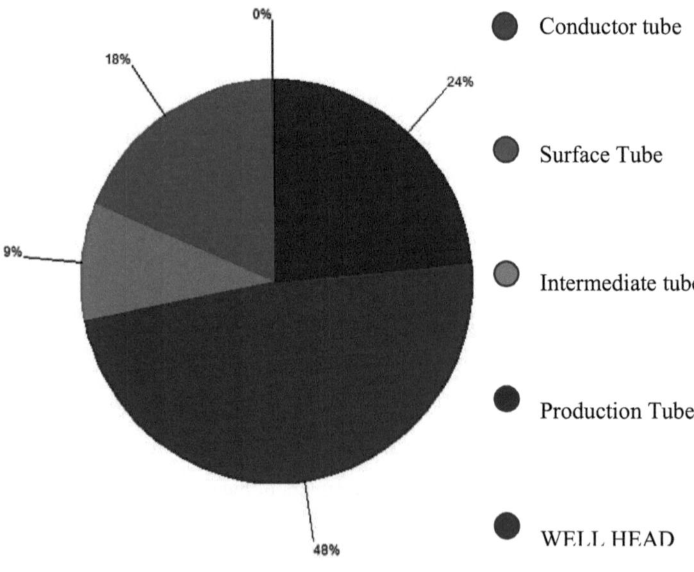

Figure 25 Probabilistic estimation of tangible costs

- **Intangible costs**

The intangible drilling costs include a wider range of categories, mainly mud services, cementing, testing, logging, fuel, etc. Like the tangible costs, these costs were assessed using the probabilistic Monte Carlo method. Assuming a 50% probability, intangible costs were estimated at $4,873,710, or 79% of total project expenditure. Figure 24 shows a breakdown of the intangible costs and the corresponding percentages.

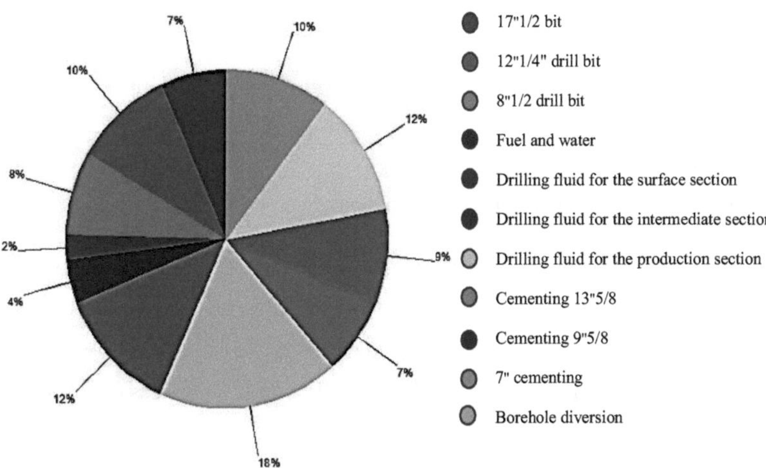

Figure 26 Probabilistic estimation of intangible costs

Par ailleurs, la profondeur du forage à un impact important sur le coût global du forage. Une représentation du coût en fonction de la progression du forage permet d'avoir un meilleur aperçu des dépenses requises à chaque phase du programme.

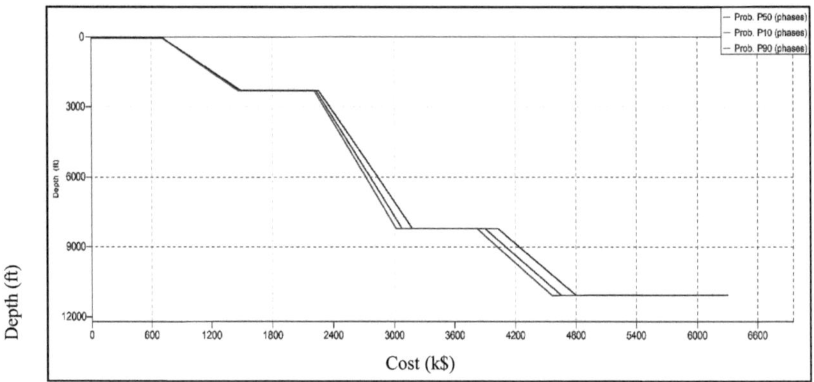

Figure 27 Trend in drilling costs as a function of vertical drilling progress

Analysis of the graph in figure 25 reveals 3 staircase-shaped curves, each corresponding to a given probability (10%, 50%, 90%). Four constant depth levels can be distinguished, corresponding to the end of each drilling phase. We can also see that the high probability curves (90%) correspond to higher costs, while the low probability curves (10%) correspond to lower drilling costs, because it is more likely to complete the project at a higher cost than at a lower cost.

The trend in expenditure on drilling the Doketi-1 well as a function of time is shown in Figure 26.

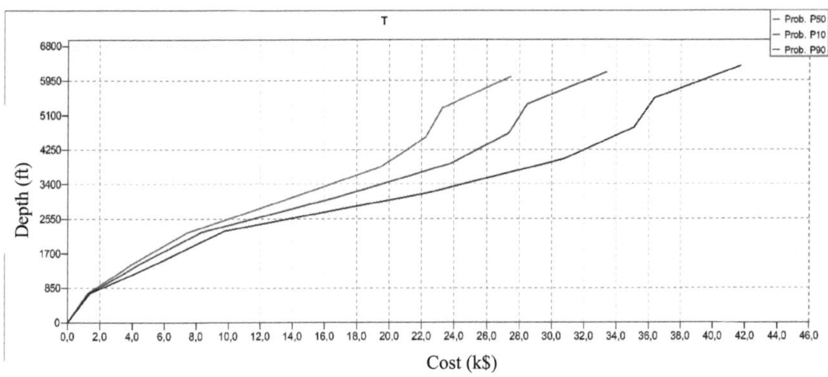

Figure 28 Graph showing the variation in expenditure over time.

The graph above shows 3 curves, each corresponding to a given probability rate: 10%, 50% and 90%. We can see a rapid increase in expenditure from day 1[er] of the project. This graph provides crucial information for planning and managing expenditure throughout the project.

The Authorization For Expenditure (AFE) is a budget document generally prepared by the operator to list the estimated expenditure for drilling. Once drawn up, this document must be provided to the partners for approval before the start of operations. This document summarises all the phases and activities that make up the programme, along with their associated durations and costs. Based on the results of this work, the average total cost of drilling the Doketi-1 well can be estimated at $6,150,216 with a 50% probability rate. Table 10 provides a summary of the costs and durations corresponding to each phase of the drilling programme.

Table 10 Summary of estimated costs for the Doketi-1 drilling programme

CLASS	DESCRIPTION	BATCON	FOSUR	TUBSUR	FOINT	TUBINT	FOROD	TEST	TUBPRO	Total
INTANGIBLE COSTS	TREPAN 17"1/2	40,000	40,000	40,000	40,000	40,000	40,000	40,000	40,000	320,000
	TRAP 12"1/4	60,000	60,000	60,000	60,000	60,000	60,000	60,000	60,000	480,000
	8"1/2 TRIPAN	50,000	50,000	50,000	50,000	50,000	50,000	50,000	50,000	400,000
	FUEL AND WATER	4,730	11,211	13,268	29,450	24,446	12,608	3,931	17,221	116,864
	DRILLING MUD SURF.	25,000	25,000	25,000	25,000	25,000	25,000	25,000	25,000	200,000
	DRILLING MUD INT.	75,000	75,000	75,000	75,000	75,000	75,000	75,000	75,000	600,000
	PRODUCTIVE DRILLING MUD.	110,000	110,000	110,000	110,000	110,000	110,000	110,00	110,000	880,000
	CEMENTING SURF.	45,000	45,000	45,000	45,000	45,000	45,000	45,000	45,000	360,000
	INT. CEMENTING	55,000	55,000	55,000	55,000	55,000	55,000	55,000	55,000	440,000
	CEMENTING PRODUC.	72,000	72,000	72,000	72,000	72,000	72,000	72,000	72,000	576,000
	WELL DIVERSION	20,270	48,046	56,863	126,213	104,767	54,034	16,846	73,805	500,844
TOTAL INTANGIBLE COSTS ($)		556,999	591,257	602,131	687,663	661,213	598,642	552,77	623,026	**4,873,708**
TANGIBLE COSTS	CONDUCTOR PIPE	600	600	600	600	600	600	600	600	4,800
	SURFACE TUBE	29,376	29,376	29,376	29,376	29,376	29,376	29,376	29,376	235,008
	INTERMEDIATE TUBE	15,063	15,063	15,063	15,063	15,063	15,063	15,063	15,063	120,500
	PRODUCTION TUBE	77,025	77,025	77,025	77,025	77,025	77,025	77,025	77,025	616,200
	WELL HEAD	37,500	37,500	37,500	37,500	37,500	37,500	37,500	37,500	300,000
TOTAL TANGIBLE COSTS ($)		159,564	159,564	159,564	159,564	159,564	159,564	159,564	159,564	**1,276,508**
TOTAL ($)		716,563	750,820	761,694	847,226	820,777	758,206	712,34	782,589	**6,150,216**

Conclusion

The aim of this chapter was to present the results obtained by applying the methods and techniques described in the previous chapter. The results show that designing a drilling programme at minimum cost necessarily involves a number of well-defined stages, including rig sizing, drilling trajectory design, casing sizing, design of the drilling mud programme, selection of drill bits and downhole assemblies, design of the well control programme, and estimation of drilling time and cost. However, for the purposes of this work, the drilling programme was designed to achieve minimum cost while complying with the safety measures set out in the Well Proposal.

GENERAL CONCLUSION

The objective here was to design a reliable deviated drilling programme for the Doketi-1 (A4-1) well in the Doba Basin that meets CNPC (China National Petroleum Corporation) standards, and to draw up its budget. However, drilling planning is a complex task that requires multidisciplinary skills. In order to develop this drilling programme more effectively, a study of each component of the drilling system was carried out. In order to ensure the safety of workers and guarantee optimum drilling performance, the following components were selected: a 1,290.6 HP winch, 3 triplex mud pumps with an individual power of 1,102.7 HP, a rotation power of 1,456.89 HP and a 45.5 m high derrick.

The drilling programme itself has been subdivided into 5 sub-programmes and the characteristics of each sub-programme have been selected as follows:

The casing programme: The conductor pipe (diameter 20", grade H-40, shoe depth 213.36 ft), the surface casing (diameter 13 3/8", grade K-55, shoe depth 2,296.6 ft), the intermediate casing (diameter 9 5/8", grade N-80, shoe depth 8,208.7 ft) and the production casing (diameter 7", grade N-80, shoe depth 11,056.4 ft).

Cementing programme: Class G cement with a density of 1.85 g/cm^3 for the surface section, class G cement with a density of 1.60 g/cm^3 for the intermediate phase and class G cement with a density of 1.50 g/cm^3 for the production section.

The slurry programme: water circulation for the conductive tube, a water-based slurry with a density of between 1.03 and 1.08 g/cm^3 for the surface section, a water-based slurry with a density of between 1.08 and 1.2 g/cm3 for the intermediate phase and a water-based slurry with a density of between 1.2 and 1.3 g/cm^3 for the production section.

The tool programme: A TMT type tool with a diameter of 17 1/2" for the surface section, a PDC type tool with a diameter of 12 1/4" for the intermediate section and a PDC type tool with a diameter of 8 1/2" for the production section.

Well control programme: A well blow-out preventer (BOP) with a working pressure of 5,200 psi.

In addition, an estimate of the duration and cost of drilling was made with a view to drawing up an expenditure authorisation. This showed that the drilling of the Doketi-1 (A4-

1) well would take 33.39 days at a total cost of $6,150,216, compared with wells generally drilled in Nigeria under almost similar conditions, which cost $25,000,000 for a total duration of 83 days.

BIBLIOGRAPHY

Azar J. (2004). Oil and natural gas drilling. *Encyclopedia of Energy*, Vol. 4, p. 35-58.

Baaziz A. (2014). The patent information for the oil and gas industry: Case study of drill bits design and optimization by reverse. *Journal of Information Systems and Technology Management*, Vol. 11, p. 645-671.

Enriquez R. (2019). Developing novel high performance to drilling mud for applications in high pressure and high temperature oil. Thesis, Durham University, 144 p.

Farah O. (2013). Directional well design, trajectory and survey calculations, with a case study in Fiale, Asal rift, Djibouti. *United Nations University*, No. 27 pp. 627-634.

Faridah S. (2021). A technical review on newly formulated. Thesis, Baze University, 143 p.

Ferreira L. (2018). Development of the enhanced directional difficulty index. Thesis, University of Lisbon, 86 p.

Hossain E. (2015). Drilling costs estimation for hydrocarbon wells. *Journal of Sustainable Energy Engineering*, Vol. 3, pp. 12-22.

Hussein A. and Fahad A. (2020). Optimization of deviated and horizontal wells trajectories and profiles in Rumaila oilfield. *Journal of Petroleum Research and Studies*, Vol. 10, N° 2, p. 19-32.

Kilian D. (2007). Casing design for deep wells in the Vienna basin. Thesis, Mining University Leoben, 79 p.

King G. (2020). Introduction to petroleum and natural gas engineering. E-Education [Online]. [Accessed 18/06/2022]. Available from https://www.e-education.psu.edu/png301/sites/www.e-education.psu.edu.png301/files.

Kruszemski M. (2017). Slimhole well casing design for high-temperature geothermal exploration and reservoir assessment. Thesis, University of Stavanger, 84 p.

Lavoisy O. (2022). First oil well. Encyclopaedia Universalis [Online]. [Accessed 21/06/2022]. Available at https://www.universalis.fr/encyclopedie/premier-puits-de-petrole.

Louis P. (1970). Contribution géophysique a la connaissance géologique du bassin du Lac Tchad. Report, Office de la Recherche Scientifique et Technique d'Outre-Mer, 355 p.

Mele K. (2021). Nigeria's top problem is well cost. Africa Oil and Gas Report [Online]. [Accessed 26/06/2022]. Available at https://africaoilgasreport.com/2021/04/partner-content.

Neal A. (2006). Petroleum engineering handbook. A complete Well Planning Approach. Oklahoma: PennWell Publishing Company. 719 p.

Nedilik G. (2021). Drilling fluid and cement slurry design. *Applied Sciences*, p. 7-14.

Nelson E. (2012). Well cementing fundamentals. *Oilfield Review*, Vol. 24, p. 18-35.

Samba P. (2018). Cours de well control. School of Geology and Mining of the University of Ngaoundéré, 70 P.

Wolfgang P. (2016). Drilling engineering. Thesis, Curtin University of Technology, 282 p.

Zenabou L. (2018). Well design: a case of KTM Z well in Douala oil field. Dissertation, School of Geology and Mining Engineering of the University of Ngaoundere, 44 p.

APPENDICES

Appendix 1: Bit classification table

<table>
<tr><th colspan="2">1</th><th colspan="2">2</th><th colspan="7">3</th><th colspan="2">4</th></tr>
<tr><th colspan="2" rowspan="2">Series</th><th colspan="2" rowspan="2">Type of training</th><th colspan="7">Bearing/Seal type</th><th colspan="2" rowspan="2">Additional features</th></tr>
<tr><th>1</th><th>2</th><th>3</th><th>4</th><th>5</th><th>6</th><th>7</th></tr>
<tr><td rowspan="12">Steel teeth</td><td rowspan="4">1</td><td rowspan="4">Tender</td><td>1</td><td rowspan="32">Standard roller bearing</td><td rowspan="32">Standard air-cooled roller bearing</td><td rowspan="32">Standard roller bearing with gauge protection</td><td rowspan="32">Sealed roller bearing</td><td rowspan="32">Sealed roller bearing with gauge protection</td><td rowspan="32">Sealed roller bearing with gauge protection</td><td rowspan="32">Sealed friction bearing with gauge protection</td><td rowspan="2">A</td><td rowspan="2">Pneumatic application,</td></tr>
<tr><td>2</td></tr>
<tr><td>3</td><td rowspan="2">B</td><td rowspan="2">Special bearing seal</td></tr>
<tr><td>4</td></tr>
<tr><td rowspan="4">2</td><td rowspan="4">Medium</td><td>1</td><td rowspan="2">C</td><td rowspan="2">Central jet</td></tr>
<tr><td>2</td></tr>
<tr><td>3</td><td rowspan="2">D</td><td rowspan="2">Deviation control</td></tr>
<tr><td>4</td></tr>
<tr><td rowspan="4">3</td><td rowspan="4">Competent</td><td>1</td><td rowspan="2">E</td><td rowspan="2">Extended jet</td></tr>
<tr><td>2</td></tr>
<tr><td>3</td><td rowspan="2">G</td><td rowspan="2">Additional gauge</td></tr>
<tr><td>4</td></tr>
<tr><td rowspan="20">Tungsten carbide teeth</td><td rowspan="4">4</td><td rowspan="4">Tender</td><td>1</td><td rowspan="2">H</td><td rowspan="2">Application for deviation</td></tr>
<tr><td>2</td></tr>
<tr><td>3</td><td rowspan="2">J</td><td rowspan="2">Jet deflection</td></tr>
<tr><td>4</td></tr>
<tr><td rowspan="4">5</td><td rowspan="4">Soft to medium</td><td>1</td><td rowspan="2">L</td><td rowspan="2">Bearing very close to the gauge diameter</td></tr>
<tr><td>2</td></tr>
<tr><td>3</td><td rowspan="2">M</td><td rowspan="2">Downhole motor application</td></tr>
<tr><td>4</td></tr>
<tr><td rowspan="4">6</td><td rowspan="4">Medium</td><td>1</td><td rowspan="2">S</td><td rowspan="2">Standard steel tooth model</td></tr>
<tr><td>2</td></tr>
<tr><td>3</td><td rowspan="2">T</td><td rowspan="2">Double cone drill bit</td></tr>
<tr><td>4</td></tr>
<tr><td rowspan="4">7</td><td rowspan="4">Competent</td><td>1</td><td rowspan="2">W</td><td rowspan="2">Improved cutting structure</td></tr>
<tr><td>2</td></tr>
<tr><td>3</td><td rowspan="2">X</td><td rowspan="2">Chisel teeth</td></tr>
<tr><td>4</td></tr>
<tr><td rowspan="4">8</td><td rowspan="4">Very competent</td><td>1</td><td rowspan="2">Y</td><td rowspan="2">Conical teeth</td></tr>
<tr><td>2</td></tr>
<tr><td>3</td><td rowspan="2">Z</td><td rowspan="2">Other tooth shapes</td></tr>
<tr><td>4</td></tr>
</table>

Appendix 2: API bit selection and casing model (Kruszemski, 2017)

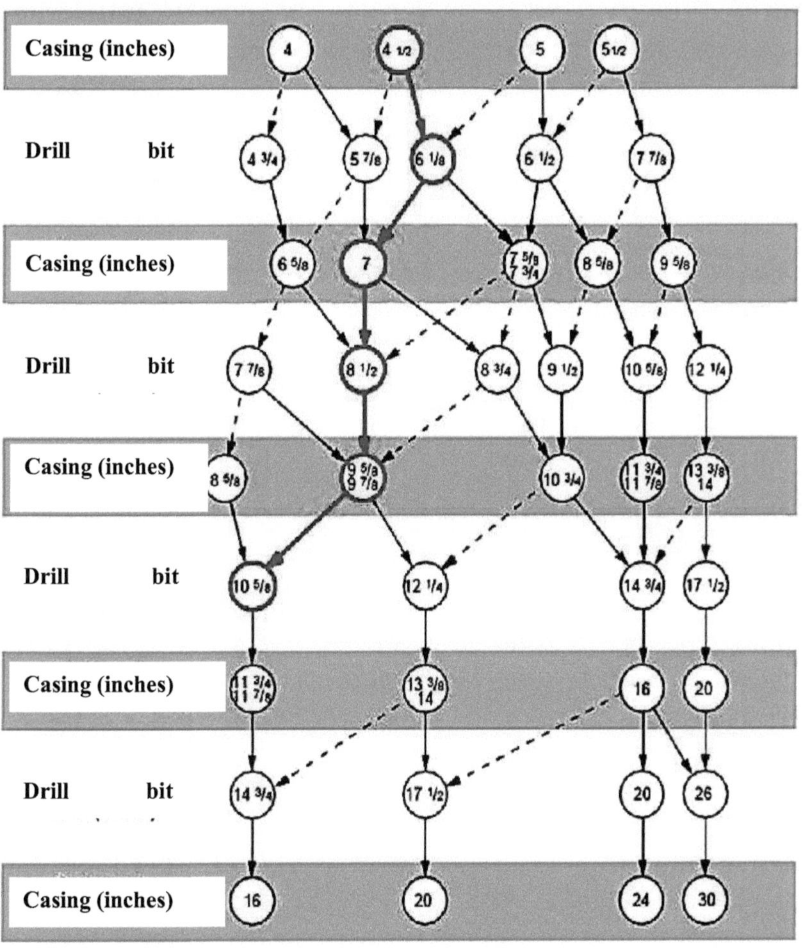

Appendix 3: Geometric data for the well trajectory

MD (ft)	INC (°)	AZ (°)	TVD (ft)	DLS (°/100ft)	Deviation (ft)
200	0	360	200	0	0
1000	0	360	1000	0	0
1700	0	360	1700	0	0
2300	0	360	2300	0	0
3000	0	360	3000	0	0
3700	0	360	3700	0	0
4500	0	360	4500	0	0
5300	0	360	5300	0	0
5900	0	360	5900	0	0
6779	0	360	6779,2	0	0
7300	15,62	169,02	7293,6	3	70,6
7700	27,62	169,02	7664,7	3	217,7
8100	39,04	169,02	7997,3	0	438,7
8600	39,04	169,02	8385,6	0	753,6
9200	39,04	169,02	8851,6	0	1131,6
9800	39,04	169,02	9317,7	0	1509,5
10300	39,04	169,02	9706	0	1824,4
10800	39,04	169,02	10094,4	0	2139,3
11111,4	39,04	169,02	10336,3	0	2335,5

Appendix 4: Basic design data

<table>
<tr><td rowspan="5">Basic data</td><td>Name</td><td>Doketi-1 A4</td><td colspan="2">Bypass</td><td>Type of well</td><td>Exploration wells</td></tr>
<tr><td colspan="4">Ground elevation</td><td colspan="2">375m</td></tr>
<tr><td>Contact details</td><td colspan="5">E: 341669.43m; N: 997848.50m</td></tr>
<tr><td>RKB</td><td colspan="5">9.0</td></tr>
<tr><td rowspan="2">Targets</td><td colspan="5">Kedeni Fm, Mangara Fm</td></tr>
<tr><td rowspan="10">Directional data</td><td>Training</td><td colspan="2">Target coordinates (m)</td><td>Target depth (m)</td><td>Proposed depth (m)</td></tr>
<tr><td>Surface</td><td></td><td>341669.43</td><td>997848.5</td><td></td><td></td></tr>
<tr><td>Target 1</td><td>Kedeni</td><td>341700.06</td><td>997690.68</td><td>2470.92</td><td>2504.32</td></tr>
<tr><td>Targets 2</td><td>Mangara Fm</td><td>341741.06</td><td>997479.71</td><td>2754.6</td><td>2860.28</td></tr>
<tr><td>TD</td><td></td><td>341802.57</td><td>997163.55</td><td>3150.49</td><td>3370</td></tr>
<tr><td rowspan="5">Tilt rate</td><td>Section (m)</td><td>Rate of Over All Angle Change</td><td colspan="3" rowspan="5"></td></tr>
<tr><td>0~1000</td><td>≤ 1.6°/30m</td></tr>
<tr><td>>1000~2000</td><td>≤ 2.0°/30m</td></tr>
<tr><td>>2000~3000</td><td>≤ 3.0°/30m</td></tr>
<tr><td>>3000~3370</td><td>≤ 4.0°/30m</td></tr>
</table>

Printed by Books on Demand GmbH, Norderstedt / Germany